Animals and Science

Animals and Science: From Colonial Encounters to the Biotech Industry

Edited by

Maggie Bolton and Cathrine Degnen

CAMBRIDGE
SCHOLARS
PUBLISHING

Animals and Science:
From Colonial Encounters to the Biotech Industry,
Edited by Maggie Bolton and Cathrine Degnen

This book first published 2010

Cambridge Scholars Publishing

12 Back Chapman Street, Newcastle upon Tyne, NE6 2XX, UK

British Library Cataloguing in Publication Data
A catalogue record for this book is available from the British Library

ISBN (10): 1-4438-2556-5, ISBN (13): 978-1-4438-2556-6

TABLE OF CONTENTS

LIST OF ILLUSTRATIONS

ACKNOWLEDGEMENTS

We wish to express our thanks firstly to the ESRC for funding the *Animals and Science: Anthropological Approaches* workshop in June 2005, from which the chapters presented in this volume have emerged, as part of the fellowship Maggie Bolton held at Manchester from 2003-2006 (Award reference RES-000-27-0026).

We also would like to extend thanks to the University of Manchester for hosting the conference and for providing us with administrative and web-site support. We received much help and encouragement from the Department of Social Anthropology at Manchester, and owe a debt of thanks to Penny Harvey, who co-organised the conference with Maggie Bolton and acted as a discussant at the event, and to Sarah Green who, having been unable to attend the conference, by way of recompense agreed to write the afterword to the volume. Thanks are likewise due to Henry McGhie for the tour of the Manchester Museum he hosted on the workshop's first evening. We also want to thank all participants at the workshop, including the contributors to this volume, and especially Sarah Franklin and Harriet Ritvo who interrupted busy schedules in order to attend and whose participation greatly enhanced our discussions. We are also much indebted to Thomas Grisaffi and Julie Wood, then postgraduates in the social anthropology department at Manchester, who gave up their time to provide valuable administrative assistance during the course of the event.

During the preparation of the volume, various colleagues have given us support, particularly through reading and commenting upon sections of the text. We greatly appreciate the time and contributions we have received from Julie Cruikshank, Penny Harvey, James Leach and Peter Phillimore in particular and more generally from our respective academic departments, the Department of Anthropology at the University of Aberdeen and Sociology at Newcastle University. Thanks also to the team at Cambridge Scholars Publications for their support and patience in the preparation of the volume.

A book like this one that seeks to reframe the relationship between anthropology, science and animals would arguably be remiss without also thanking the non-human animals in our lives. To Phoebe, the late Boris, Tigger and Dibley: thank your for offering unstinting solace when needed and endless hours of shared contentment, punctuated at times by relentless pestering when our preoccupation with the computer meant that the food dish was in danger of becoming (unacceptably) empty. Sméagol who was not around for most of the volume's preparation, has since his recent arrival taken an uncommon interest in cursor movements on the computer screen and has keyed in some suggestions of his own for editorial changes to the volume's text.

—Maggie Bolton and Cathrine Degnen

INTRODUCTION

ANIMALS AND SCIENCE: FROM COLONIAL ENCOUNTERS TO THE BIOTECH INDUSTRY

MAGGIE BOLTON AND CATHRINE DEGNEN

Sirius: An Early Step into the World of Animals and Science

On reviewing some of the contributions to this volume, one of the editors was reminded of a work of science fiction she enjoyed as a teenager: a little-known novel called *Sirius: A Fantasy of Love and Discord*, by Olaf Stapledon, a published philosopher and lecturer at the University of Liverpool (Stapledon 1944). In a way, *Sirius* is a reworking of the Frankenstein story, as it starts with a scientist, Thomas Trelone, who creates a creature by artificial means. Rather than produce a monster of human form, the scientist of this story succeeds in producing super-intelligent dogs through introducing hormones into the bloodstreams of pregnant bitches. The book was written long before either cloning or genetic modification had entered the popular imagination, let alone become achievable realities. Trelone's first efforts produce enhanced border collies, "super-sheepdogs" prized by the hill farmers near his family home in Wales: dogs that are efficient workers, but still unambiguously canine. His life's work culminates with the birth of Sirius, a puppy who surpasses all his previous efforts and proves to have not only the intelligence, but also the pattern of growth and development of a human being.

Although the novel gives Stapledon a chance to explore what to human beings are acceptable and unacceptable outcomes of science, it passes overt moral judgment neither on the science of creating this anomaly nor on the individual scientist. Instead, through dwelling on Sirius's existential angst, Stapledon provides us with an exploration of what it means to be human, and of the boundaries between humans and animals. The dog is torn between his yearnings to partake of human intellectual pursuits and pursue human relationships and his non-human "wolf-nature" that leads him to hunt, kill savagely and chase bitches.

The character of Sirius gives Stapledon the opportunity to comment on human relationships, prejudices and activities from a position outside the human species. Individual human characters display different reactions when faced with a transgression of the human/animal boundary and the resultant confusion of categories that Sirius represents. Trelone's daughter, Plaxy, with whom Sirius is raised and has an intense relationship, is torn between her love for him and resentment that her closeness to him has made her distant from her own species. Working relationships with Trelone's Cambridge colleagues are difficult, for Sirius at one time harbours hopes for a scientific career of his own, yet as an animal is an object of scientific inquiry for others.

Sirius is both like us and unlike us–as Mary Midgley (1994:36) notes, the term animal can be used both in inclusive and exclusive senses: as a category that includes the human and as the unhuman or anti-human. In the latter sense, "animal" becomes a symbol for the forces we fear in the world around us, that we regard as alien to ourselves. However, she comments, "the peculiar alarm they produce suggests that they are not altogether alien, that we too carry the seeds of them within our nature" (*Ibid.* 35). It is thus not surprising that the story ends tragically. Sirius encounters outright hostility from conservative villagers who have grown suspicious about the moral propriety of his relationship with Plaxy and from a non-conformist pastor convinced that the dog is possessed by Satan. Sirius, who has always regarded hunting and killing as part of his "wolf-nature", himself becomes hunted as he is pursued across the Welsh mountains and shot dead by humans.

We returned to *Sirius* while preparing this collection of essays from anthropology and related disciplines that focus on interfaces between animals and science. Stapledon published his novel almost twenty years before Claude Lévi-Strauss declared that animals are "good to think"[1], and seems not to have anticipated the work of the intervening generation of anthropologists eager to use animals to explore particular human social formations. He seems instead to be more in tune with the recent "animal turn" in the social sciences that rejects such symbolic reductionism and sees animal-human relationships in themselves as worthy subjects of study. *Sirius* also has clear resonances with a recent "turn to dogs" in scholarship at the intersections of animals with science (Haraway 2003, 2008; Smuts 1999). Although the essays included in this volume deal with recent scientific innovations and address debates that simply did not exist half a century earlier, Stapledon's main concerns resurface in their pages. The contributions to this volume look at what it means to be human, the place of human beings *vis à vis* other species on this planet, our ideas of what is nature and what is culture, the limits to our ideas of kinship, the ethical debates that surround science along with their interpretation by both scientific communities and lay publics, and the moral comportment of scientists. They also add a dimension that Stapledon was unable to address in so short a novel: that of comparison between the knowledge practices concerning animals of people in different ethnographic settings and our own to which Western science always forms a backdrop.

Genesis of the Volume

This volume grew out of a three-day workshop held at the University of Manchester in June 2005 that Maggie Bolton organised together with Penny Harvey. The workshop was financed by the ESRC as part of a research fellowship Bolton held at Manchester at the time, during the course of which she had carried out an ethnographic study of llama herders in the Bolivian Andes faced with a scientifically informed livestock development project (see Bolton, this volume). This study had made her interested not just in how animals figure in scientific discourse, but also in the ways that scientific innovations work to reconfigure relations between humans and other species. She was inspired to organise the workshop by the thought of bringing together a variety of anthropologists and scholars from related social science disciplines whose work on animals, animal-human relations and science resonated with concerns that were surfacing in her own work.

During the same period of time Cathrine Degnen was likewise based at Manchester. Together with Ben Campbell, who also contributes to the volume, she was working as part of a team engaged in a large research project: a Europe-wide investigation of the public understandings of genetics, headed by Jeanette Edwards.[2] Both Degnen and Campbell were involved in a strand of this project that looked at public understandings of genetically modified food and the relationships between "non-expert" knowledge, food and genetics. Degnen decided to contribute to the workshop following a conversation with Harvey about British debates over the genetic modification of plants and crops. Harvey enquired, "Where are the animals?" and Degnen wrote her contribution to the workshop and later to the volume as a response to this question. Her participation in the workshop later led to her joining the editorial team for the publication.

The workshop was attended by scholars representing a variety of disciplines from anthropology to veterinary science. The paper presenters ranged from established researchers such as anthropologist, Sarah Franklin, and cultural historian, Harriet Ritvo, presenting chapters from publications already in the advanced stages of planning (see for example Franklin 2007), to recent PhD graduates and postdoctoral researchers. An event that set the tone for the workshop on the evening of the first day was a tour of some of the animal exhibits in the Manchester Museum, given by Henry McGhie, then curator of zoology and now the museum's Head of Natural Environments. McGhie's discussions of taxidermy and taxonomy encouraged us to think through the biographies of animals to the relationships in which they had been entwined. Recalling Donna Haraway's discussion of taxidermy in the American Museum of Natural History (Haraway 1989a), his talk also illustrated to us the ways in which the remains of animals have been used to make statements about the position of humans *vis à vis* other living beings and about European society *vis à vis* the rest of the world.

Contextualising Animals in Anthropology: From Totemism to Biotechnology

Before embarking on a summary of the themes broached by this volume and how they are addressed by the different contributors, there is a need to sketch out some of the intellectual genealogy behind the perhaps unlikely combination of animals with science in a collection of essays that fall largely within the discipline of social anthropology. To start by contextualising animals in anthropology, one might note that although animal-human relations did not feature strongly in early ethnographies, anthropology has long paid considerable attention to various beasts, if not in the flesh, then through their appearances in myths and in the names of groups of humans in diverse societies.

The naming of human groups after animal species captured the attention of nineteenth- and early twentieth-century scholars who came to regard totemism as an important institution of so-called primitive societies (Durkheim 2001; Radcliffe-Brown 1952). This reification of the concept culminated in an essay by Lévi-Strauss (1963) that debunked it as a valid area of study, yet simultaneously declared that animals are "good to think" for humans (*Ibid.* 89). It should be added that Lévi-Strauss was not particularly interested in animals themselves. He was interested in them only in so far as he saw the observable differences between animal species as a vehicle that permitted the embodiment of ideas and relations conceived in abstract human thought: in particular, of offering a way to naturalise the social distinctions and classifications of human society.

The different ways that humans mobilised the differences in form and habits between species of animals led to various influential works in anthropology on the subjects of classification and boundaries (Douglas 1966; Bulmer 1967; Leach 1964). Although the structuralist thinking of the 1960s has long since fallen out of fashion, a concern with boundaries and classification, and the place of animals in relation to humans, retains an importance in anthropological debate. This is particularly the case in discussions of the cosmologies of different peoples and the cultural specificity of the division between the things of nature and the culture of humans that undergirds Western post-Enlightenment thought (e.g. Descola 2005; Descola and Pálsson (eds.) 1996). It is fair to note along with Russell (2010: 17), however, that anthropology, as a study of human beings, has tended to take the human-animal boundary itself somewhat for granted and to draw it rather starkly–perhaps rather more starkly than do people in some of the societies that anthropologists study.

While animals were clearly important to anthropology in the early twentieth century they took a predominantly passive role in anthropological accounts. They were not treated as active subjects, and their presence was contingent upon them providing (through their names, or through the consumption of their bodies) significant insights into relations between different human beings or groups of humans. Classic examples of this are Evans-Pritchard's studies of the cattle-obsessed Nuer in the Sudan (Evans-Pritchard 1940, 1956).

Nevertheless, animals in other pastoral societies entered into scholarly work in a different way. Anthropologists and archaeologists from their respective disciplinary perspectives were interested in processes of domestication, particularly of herd animals (e.g. Clutton-Brock (ed.) 1989), and in the ways that human beings in different societies envisage their relations with the beasts they tend. Some extrapolated from this to explore the consequences of particular sorts of human-animal relations not only for animals, but also for inter-human relations in the societies in question, arguing that the domain of social interactions between humans cannot be separated from the domain of interactions between humans and animals. Following fieldwork with reindeer herders in Finland, Ingold (1980) considered the relations with animals of hunters, pastoralists and ranchers and the varying degrees of domination exercised by humans in those relationships. This led him to hypothesise that the transition from hunting to herding (to which domestication was crucial) and eventually to ranching has been accompanied by a replacement of egalitarian and sharing relations, characterised by trust, with hierarchy and domination. Returning to our earlier point about the demarcation of a boundary between animals and humans, recent work coming out of both northern Aboriginal societies and Amazonia suggests that hunters not only conceptualise their relations with animals in terms of equality, but also find the human/animal boundary less stark or absolute than do members of other types of society (Willerslev 2007; Viveiros de Castro 1998).

Domestication has remained a subject of debate in anthropological studies of animals. Works on the subject written in the 1980s, reflecting theoretical trends in the discipline at the time, tended to emphasise the conversion of nature into property (Clutton-Brock (ed.) 1989). However, more recently scholars from a variety of disciplines have sought to shift the focus away from domination and property ownership, stressing instead the symbiotic and serendipitous character of processes of domestication (e.g. O'Connor 1997) and the active roles of animals themselves in entering the lives of human beings (Budiansky 1992). Ingold's view of domestication as linked to domination and the eclipse of sharing relations has been challenged, for example, by Penelope Dransart's (2002) work on Andean herding (see also her contribution to this volume) and more recently by the cross-disciplinary contributors to a collection of essays edited by Rebecca Cassidy and Molly Mullin (2007). As Donna Haraway has pointed out insightfully, stories of domestication or "co-evolution" function as powerful "origin stories" that define relationships between nature and culture and also "who and what counts as an actor" (Haraway 2003:27).

Although studies in cultural history in the 1980s started to take the roles of animals in society seriously in their own right (Ritvo 1984, 1998), a turn to animals in anthropology came somewhat later. Perhaps a precursor was Eduardo Archetti's ([1992] 1997) study of guinea pigs in Ecuador, a work initially aimed at increasing understanding of indigenous people among the Ecuadorian middle classes. Archetti makes these small domestic rodents, found scurrying around Andean kitchens, the focus of a study that culminates in conflicts of knowledge between indigenous people and developers. Although his work focuses on the symbolic dimensions of guinea-pigs, it also emphasises their embeddedness in Andean people's lives. In the years that followed, the designation of "Animal Studies" as a legitimate area of interdisciplinary academic inquiry turned studious attention further in this direction. Scholars became more interested in the details of human-animal relations as *parts* of human society rather than just *symbols* of it (Knight 2005a).

In the last decade, various studies have emerged in which animals "get to count as actors" to a much greater extent than had previously been the case. Cultural geographers Philo and Wilbert (2000), along with their contributors, explore the spatiality of animal-human relations: the co-presence of animals and humans in human spaces and the assignment of animals to particular spaces where they are supposed to remain. In anthropology, the complexities, ambivalences and intimacies of relations between humans and animals are a focus of the contributors to the collection edited by John Knight mentioned above (Knight (ed.) 2005b) and recent years have seen the publication of essays and monographs focusing on humans and their relations with beasts of different shapes and sizes in different ethnographic contexts: thoroughbred horses (Cassidy 2002), rats (Edelman 2005), captured elephants (Locke 2008 and in preparation) and water buffalo (Campbell 2005) in Nepal, llamas and alpacas in Chile and Bolivia (Dransart 2002; Bolton 2006), dogs (Haraway 2003) and meerkats (Candea 2010) to name but few.

Nevertheless, the focus on relations with animals does not mean that thinking through animals has ceased to be useful for providing significant insights into those aspects of human culture that are the conventional concerns of anthropology, such as kinship or the ownership of property. In a way that parallels recent shifts in gender and kinship studies towards science, the arrival of new reproductive technologies, together with advances in biotechnology and genomics, have led science to play a prominent role in anthropological accounts of human-animal relations. This preoccupation reflects the to-ing and fro-ing of new reproductive technologies between animals and humans in the process of their development, and revelations from DNA analysis of how very small the biological differences are between humans and other animals. In such accounts, we begin to see how meetings between animals and science are starting to set new theoretical agendas for anthropologists

Anthropological discussions of notions of property where animals are concerned have drifted away from studies of domestication and towards accounts of the biological makeup of beasts. The idea that the bodies of animals contain a heritable substance, or template for future generations, that is a form of capital considered to be the intellectual property of animal breeders, forms the basis of Ritvo's (1995) essay on early sheep improver, Robert Bakewell. Sarah Franklin's (2007) study, *Dolly Mixtures*, inspired by Dolly, the sheep cloned at the Roslin Institute in Scotland, revisits notions of "capital" and "stock", noting along the way how these terms originated with holdings of live animals. Franklin adds to current discussions of rights of ownership not only in biological materials, discussing cell lines and sequences of DNA, but also, and importantly, to discussions of innovation and intellectual property in techniques of manipulating the said materials, notably the technique of somatic cell nuclear transfer used to create Dolly. This technique, which is many times more valuable than the substance of the sheep herself, becomes the combined property of the scientists behind her gestation and of the various organisations providing capital to fund the research in the first place.

Paul Rabinow (1999: 408) (taking up a quotation used by Gilles Deleuze) is intrigued by a cryptic remark of the poet Rimbaud that has come to take on a very material meaning in this age of genomics and genetic modification: that "the man of the future will be filled (*chargé*) with animals". This remark finds echoes in Donna Haraway's recent works, *The Companion Species Manifesto* and *When Species Meet* (Haraway 2003, 2008), which develop themes that emerge from her earlier discussions of cyborgs (Haraway 1991). In these works Haraway skilfully negotiates a path between abstract philosophical theory and everyday lived relations and interactions with animals. Her intention is to highlight the mutual dependence of forms of life, one with another, and to deploy animal-human relations to emphasise the physical, social and emotional entanglement of species. She aims thereby to challenge the "human exceptionalism" (Haraway 2008: 11-12) that views humans as ontologically unique, in not being situated within a web of interspecies dependencies, and thus privileges human beings over other animals in everyday assumptions, philosophy and policy making. In her opinion, such a view has in any case been destabilised irretrievably by recent advances in biotechnology that confirm the interspecies commonality of much genetic substance. Haraway illustrates her arguments through the relationship she has developed with her dog: not only the social interactions involved in living side by side with an animal, but also the exchanges of biological material that take place when we share our lives with companion species. As a result, she argues, we and animals literally "make each other up in the flesh" (Haraway 2008: 16). Her philosophical line of reasoning leads her to join Bruno Latour in challenging a series of "Great Divides": between nature and culture; human and animal; organic and technical; and kinship and ownership, a challenge that, like Latour's work, has radical political as well as philosophical implications.

Anthropology and its Concerns with Science

Having sketched out anthropology's concerns with animals, terminating with the ways in which studies of animals have started to intersect with those of science, it remains to give a brief resumé of anthropology's relationship and concerns with science itself. Science, as both a way of knowing the world and as a set of distinctive practices that helps create meaning about the world, has increasingly become the focus of anthropological attention. This is partly due to longstanding anthropological interest in the points of intersection between knowledge, belief and practice that science makes manifest. It is also partly due to the reflexive turn in anthropology since the 1980s that has prompted greater scrutiny of knowledge practices closer to "home" and has led to sustained problematising of inherent assumptions about epistemology and ontology held within scientific models of explanation (Edwards, Harvey and Wade 2007). But science of course does not remain at arm's length from the rest of society. It is, as Franklin reminds us, "a foundational belief system" in Western societies that is also "a source of cultural values" (1995:165), but one that is simultaneously embedded in the cultural systems from which it stems. Science is thus a site of critical importance for understanding how knowledge and meaning are established in the near to hand.

In addition to such long-standing concerns with knowledge, belief and practice, anthropology has also long been concerned with examining the ways in which power is exercised and reproduced via discursive practices and epistemological regimes. Science holds a particularly powerful place in Western cosmologies and ideologies, the roots of which are in the Enlightenment, and as Nader starkly asserts, "science anchors power" (1996: 12). Furthermore, a "belief in the omnicompetence of science" in the Western world has been building momentum throughout the 20th century (Nader 1996: 24). As such, science becomes a riveting domain in which to interrogate the production and reproduction of power and knowledge in Western society. Working from within, however, presents its own sets of challenges, particularly in navigating these power relationships in which the ethnographer of science is both immersed and subordinate and so consequently "studying up" (Franklin 2002:354). Additionally, from a Latourian point of view scientific endeavour has exacerbated the purification work that divides up natural and cultural worlds and which conceptually holds science apart from culture (Nader 1996; Edwards, Harvey and Wade 2007). Working against this deeply imbedded indigenous belief is something anthropologists studying science take on. That the anthropology of science should be willing to take such challenges up, however, should not come as a surprise, given the ways in which the field has also built upon and drawn from feminist analysis of gender and kinship that threw into question the status of the biological and the social (see Franklin 1995: 168-170). This contribution, one that Franklin has termed a "gender to science shift" (1995:170), has provided fruitful theoretical ground for rethinking other cultural certainties, evidenced as she reminds us in seminal works by Strathern (1980, 1988, 1992) and Haraway (1989b, 1991, 1997).

Science has further been a productive site of ethnographic inquiry because far from being homogenous, it encompasses great diversity. The relationships between science and society are equally complex. Convincing examples of this come from Knorr Cetina (1999), on the knowledge practices of the epistemic cultures of experimental high energy physics and molecular biology, and from Traweek (1996), on power differentials amongst international science networks and within national science programmes. Anthropological approaches to science have thus also been multiple. Some anthropologists and sociologists have approached the study of science by working alongside scientists in the laboratory. The allied fields of Science and Technology Studies (STS) and Sociology of Scientific Knowledge (SSK) have both relied heavily on Actor Network Theory (initially developed by Callon 1986a and Latour 1987), a paradigm that shares many points of interest with anthropology, but which is balanced by other anthropological challenges such as "to develop translation and mediation tools for helping make visible the differences of interests, access, power, needs, desire, and philosophical perspective" (Fischer 2003:3) and attending to "the crossing-over between the culture of the lab and the culture of which this culture is a part" (Franklin 1995:174).

While the natural sciences are vast in scope, it is well established that their gaze, like all forms of knowledge, is a partially selective one: certain questions and paradigms come into focus at certain historical moments and simultaneously screen out other potential candidates (Kuhn 1970). Particular domains of concern explode into prominence and then may eventually recede as others come to supplant them. Biotechnology, genetics and genomics are current occupiers of this sort of boom in scientific attention, and Keller (1995) provides a riveting account of such shifts by focusing on transformations within biology in the twentieth century (Fischer 2003: 348-9) and as such offers a pertinent reminder of both the historicity of Western science and its malleable parameters.

As we have previously noted, science is entering into anthropological accounts of animals to an increasing extent and is thereby providing new directions for anthropological theory. However, what exactly does a focus on animals add to the anthropology of science? In this volume, we offer a response to this question, crafted through ethnographic and historical studies. Our collective work crosses disciplinary boundaries and draws from anthropology, history, sociology, and art. At stake in our discussions are a wide range of animals, sciences, ethnographic regions, and historical time frames. Collaborators on this volume have also employed very different ways of "doing anthropology", which in turn reveal different aspects of the cultural processes at work in the relationship between animals and science. Taking the chapters together, a mosaic of meaning begins to emerge that reveals particular aspects of science as practice and knowledge when non-human rather than human animals are put at the centre of inquiry. Adopting a decentring perspective, or viewing science from the margins, engenders valuable insights into foundational cultural assumptions that would take animals to be handmaidens to science rather than considering them to be co-conversationalists in how scientific knowledge comes to be what it is in the first place.

Themes Addressed in the Volume

Knowledge Practices, Classification and Boundaries

One of the most striking points of intersection between animals and science in the chapters of this volume is in regards to competing epistemologies and forms of practice. Stephanie Bunn, working in Central Asia with Kyrgyz pastoralists, examines this rubric through Kyrgyz relations with and knowledge of animals. Through the lens of food, Bunn contrasts ways of knowing and forms of evidence premised on Western science with those of the Kyrgyz people she worked with, and poses the pertinent question: how can such considerations recalibrate the relationship between science and animals? She argues that extensive practical and empirical knowledge is held by Kyrgyz people about the properties of the food they produce and eat from their animals (such as medicinal properties), and that food holds qualities that heal, protect and are attuned to unfolding social circumstances. Rather than attribute this to "ritual belief", Bunn shows how it is part and parcel of a paradigm of nature in which relationships between supernatural powers, animals and the environment are integrated into, a specific "philosophy of nature" as she terms it. Bunn demonstrates how Kyrgyz knowledge of animal and human relations are not just practical or symbolic, but intersubjective and affective, lessons that she would like to see extended to Western science.

We return to the possibilities she puts forward for this extension below in the section "Nature and Culture".

Marcia Stephenson engages in a similar set of considerations. Like Bunn, she is dealing with the paradox of two different bodies of knowledge. Drawing from Sawday (1995), Stephenson explicitly names them as the *physical* animal body and the body of *knowledge produced* about animals. While Bunn links these two bodies of knowledge through food, Stephenson shows how they are tied together via dissection. Her account is grounded in 18th century European "picturesque anatomy" of camelids and colonial expansion in the Andes. She demonstrates the ways in which anatomy and dissection of these animals (novel curiosities to a European audience) are enmeshed with the acquisition of knowledge, economic gain and exploitative colonial encounters. While dissection was practised by Andeans and Europeans alike, the knowledge gained by the colonisers' cutting and mapping of the interior of the camelids' bodies simultaneously erased indigenous knowledge and appropriated the animals as an economic resource to be profitably exploited and made "useful", particularly for their fleece.

Maggie Bolton and Penny Dransart share an ethnographic interest with Stephenson in Andean camelid pastoralists, and again, both authors contribute to this theme of competing epistemologies and practice. In Bolton's case, this is based on contemporary fieldwork in Bolivia with llama herders who are under pressure by NGO veterinarians and agronomists to administer anti-parasitic drugs to their animals. The herders however are not motivated by the same (economic) rationales as the development NGO scientific professionals. This divergence reproduces social and class tensions as well as revealing a substantial gap between Andean and Western concepts, and the persistently colonizing perspectives of science as practised by development NGOs in this context. For her part, Dransart draws on a series of comparisons between camelid herders in the Andes (Isluga, Chile) and European (Welsh) camelid fanciers. Dransart then maps the divergences in herding practices in these two cultural and geographical locations including different responses to properties of herd animals (such as leadership), different possibilities in the characteristics of relations between humans and camelids, differences in categorisation and different understandings of handling and herding practices. Such differences in herding practices, as Dransart demonstrates, are grounded in divergent understandings of autonomy, notions of property and inter-agential relations between humans and herd animals. Radical divergences in received knowledge about classificatory schema of human and non-human animals in turn develop radically different priorities in everyday practice.

Attempts at classification on such contested terrain unsurprisingly can lead to dispute. This is evident in cross-cultural contexts as the examples above ably illustrate, but as the chapters by Whitehouse, Degnen and Nimmo show, classifying and ordering in relation to animals and science present real dilemmas when contained within Western settings, too. Throughout this volume, multiple forms of ordering and classifying are evident, made possible by techniques and practices discussed here already as widespread as eating, herding, tending, "development" and anatomical aesthetics, but also via conservationism, sport shooting, biotechnology, microbiology and bovine IVF.

Nature and Culture

Another theme that is broached repeatedly in this volume is that of the "Great Divides" of post-Enlightenment thought: between nature and culture or between human and animal. Over the years anthropologists have shifted from writing from within this tradition and assuming such divisions as universals of human thought (Lévi-Strauss 1992, and see also Ortner 1974), to pointing out the cultural specificity of both nature and culture, and their non-applicability to non-Western societies (Strathern and MacCormack (eds.) 1980). More recently several have been concerned with looking at ways in which post-Enlightenment assumptions underlie thinking both in the sciences and social sciences and the ways that they impact on people from different intellectual traditions in colonial and post-colonial encounters (e.g. Shepherd 2006; Verran 2002). In the volume, several contributors use the intersection of animals with science to throw light either on changing understandings of the nature/culture or human/non-human divide, or to provide powerful critiques of existing theorisation around the topic. It is therefore perhaps unsurprising that an interlocutor to a number of the chapters is science studies scholar, Bruno Latour, particularly through his unmasking of the "Modern Constitution", the sleight of hand through which the things of nature are kept apart from the politics and culture of human beings (Latour 1993). Latour also features through his efforts to undo the great divide and reconfigure both science and social science in the process (Latour 2005) and through his blueprint for remaking environmental politics (Latour 2004).

Before commenting on those chapters that engage directly with Latour's work, a further mention of Stephanie Bunn's contribution is appropriate for, although her work stems from an intellectual tradition outside science studies, Bunn's conclusions have a strong intersection with Latour's work. Through her consideration of Kyrgyz animal knowledge,

Bunn suggests that science should find a place for the sort of detailed local and other knowledges of non-Western people. However, the reconfiguration she suggests does not stop with science, but extends to anthropology. Anthropology, she argues, has made too sharp a distinction between culture and nature, or between what it considers symbolic or metaphoric–with which it is concerned–and practical, mundane knowledge derived from sensory and intersubjective experience–which it has frequently ignored. Such a distinction, she argues, limits our ability to engage with the intellectual landscape of others, which should be one of our goals. It is hard not to identify resonances here with Latour's project of "reassembling the social", his critique of the social sciences in general. The social sciences, he argues, assume "the social" to be a special domain governed by nebulous and ill-defined forces, like "political power". He would rather the social sciences were reconfigured more as the "tracing of associations" (Latour 2005: 5), the connections between things not social in themselves, to show how such associations give rise to effects like power. However, in his earlier work Latour appeared to consider anthropology, with its ability to deal calmly "with the seamless fabric of what I shall call 'nature-culture'", an example that other social sciences should follow (Latour 1993: 7). A consideration of the animal knowledge of a non-Western people appears to have brought Bunn to similar conclusions as Latour, but without his idealisation of anthropology, towards which her critical gaze is turned.

Moving on to the contributions that directly engage with Latour's work, it is the notion of the "Modern Constitution" that sociologist Richie Nimmo addresses when he takes as his subject the diseases that pass between animals and humans. The anxiety surrounding such diseases has become a familiar feature in these early years of the twenty-first century following outbreaks of avian and swine 'flu and the prominent coverage they have received in the media. However, Nimmo's study is based not on present-day epidemics, or epizootics,[3] but on past anxieties over animal-borne disease. Nimmo takes as his focus an outbreak of rinderpest or "cattle plague" in the British dairy industry of the nineteenth century together with subsequent efforts to render the bodies of animals safe for humans to consume and uses this outbreak of disease to argue that animal-human relations have been absolutely central to defining the boundaries of the human in late-modern thought. In making this assertion he critiques Latour's formulation of the "Modern Constitution" for relegating animals to the sidelines of the nature/culture divide and giving precedence to things–especially technological objects–over beasts.

Rather than "Modern Constitution", Nimmo prefers to use the term "humanism", in emphasising the species relations underpinning the human/non-human divide of late-modern ontology. His use of the term recalls not so much its use in some postcolonial writing–in which it refers to attempts to posit the unity of humankind and thereby subsume the non-Western other to a Western ideal of the human[4]–but something approximating Haraway's "human exceptionalism" (see above): an axiomatic assumption that human beings inhabit an ontologically unique domain. Both Nimmo and Haraway are interested in connections between species, animal and human, but while Haraway (2008) has looked at how humanism (or human exceptionalism) at this present time is being undermined by interspecies meetings and advances in biology, Nimmo looks to the past–to how it became entrenched in modern thought and its materialisation in a complex of everyday practices and institutions. Such everyday practices included in the nineteenth century the efforts required to render liquid milk a safe and clean product for humans to consume–the subject of Nimmo's historical research. As milk was rendered safe and pure, he argues, the boundary between humans and animals was simultaneously "purified" in the Latourian sense, with humans coming to occupy a unique domain incommensurable with the non-human world.

Making the bodies of animals safe for humans to consume is also central to the contribution by Maggie Bolton. As noted above, the focus of this study is an attempt by development professionals in Bolivia to persuade llama herders to administer veterinary medicines to their animals amidst the racial politics of the Andean region. Developers hope that improving animal health and hygiene will help transform llama meat into a food of choice for urban consumers and as a consequence that the commercial possibilities of llama herding will be enhanced, thereby improving the livelihoods of rural people.

Bolton examines how the application of science, as "universal knowledge", can attempt to reconfigure relations between humans and animals. At its core is the attempted transformation of an animal crucial to Andean sacrificial ritual, and the maintenance of relations between humans and supernatural beings, into a commodity: a transformation regarded with suspicion by herders. Bolton acknowledges the usefulness of actor-network theory (ANT), the approach developed by Bruno Latour, Michel Callon and others, as an approach to understand the entanglement of what are conventionally held to be science and politics, in unravelling the complexities of what at first sight looks to be a straight-forward application of animal medicine. The approach also illuminates how particular configurations of human and non-human actors give rise to what we call power. However, she considers the approach to have its limitations, and its proponents to find it difficult to avoid the assumptions of post-Enlightenment thought that they wish to critique. Like the development professionals in the Andes, ANT theorists assume the various actants they posit to have "interests", which all too often seem to be those of "rational economic man" (see for example Callon 1986b). Herders however do not always behave as rational economic actors and at times prioritise the maintenance of relationships over furthering interests of their own.

Nature and culture in the cosmology of contemporary Europeans come under scrutiny in Ben Campbell's chapter. Genetic modification disrupts the biological discreteness of humans and other organisms by distributing parts of previous wholes in new combinations. Campbell uses the public debates into genetic modification that took place in the UK in 2003 to suggest that biotechnology has destabilised the modernist division between nature and culture, and between human and non-human. One specific aim of his chapter is to interrogate Descola's (2005) characterisation of the cosmology of contemporary European society as "modern naturalism" divided between an objective material world of the non-human, knowable through science, and a subjective realm of human meaning.

While opposition to genetically modified (GM) crops based on their (real or imagined) threat to the environment is sometimes phrased in terms that recall property relations (heritage, for example), Campbell suggests we understand it through a different idiom. For him, the new types of knowledge and practice represented by GM organisms have made visible the sort of relationships between humans and non-humans that we normally associate with non-Western cultures: in spite of Descola's modernist cosmology, modern people understand themselves to be related to and share citizenship with wildlife. While this conclusion brings to mind Latour's (2004) template for a public politics of associations between humans and non-humans, Campbell sees the relationships between humans and wildlife as of a more intimate nature. He builds upon Donna Haraway's (1997, 2008) arguments against human exceptionalism, and suggests that in a post-humanist, or post-exceptionalist, world we could usefully employ the notion of kinship to describe such animal-human relationships, and that we should rethink what we mean when we talk about a "natural relation".

Possession and Ownership

The theme of possession and how it relates science to animals surfaces in several chapters in the volume. We have already pointed out the etymological relation between "capital" and cattle, and "stock" and "livestock" that Franklin (2007) noted, and here we further note the common association of both the scientific revolution and the theorising of property within the same historical period. In seventeenth-century England, as the scientific revolution consolidated nature as a domain separate from the politics and society of humans (Shapin and Schaffer 1985), John Locke set out in his second treatise on government how human beings come to claim ownership over the things of nature, including animals, through the performance of labour upon them (Locke [1690] 1988).

Marcia Stephenson's contribution focuses on a historical period shortly after Locke's time: the eighteenth century and the height of the European Enlightenment. As Mary Louise Pratt (1992) has noted, this was a period when the systematisation of nature was a global project, and when knowledge was sought through the opening up of interiors, whether interiors of continents, or, as in the cases documented by Stephenson, the bodies of animals. The collection and dissection of camelid specimens related in Stephenson's chapter unfold against the background of a new phase of colonialism in which European powers competed to acquire sources of raw materials from their overseas territories. For Stephenson, the actions of natural historians and anatomists also form a metaphor for the dismemberment of Andean traditions and ways of living at the hands of Europeans.

Penelope Dransart's chapter on camelids also addresses themes of possession and property. In Locke's formulation, the things of the earth, including the "beasts it feeds" are given by God to mankind in common. They may be removed from that common state by the application of labour. Animals are part of the commons (or nature) and do not get to have property of their own. Dransart makes the point that notions of property in turn influence what sort of characteristics (or properties) are attributed to animals. Western possessive individualism does not only infect people in Western societies who own animals, but also, as she notes, the anthropological accounts of animal ownership written from within the Western tradition, that envisage animals as alienable property of humans. Such accounts do not generally see animals as exercising control over their bodies or the products of their labour, nor as having the property of being themselves acting subjects "empowered to care" (c.f. Ingold 1994).

On the basis of fieldwork in Isluga in northern Chile, Dransart is prompted to wonder about the social relations in which animals participate there and whether they might be able to reciprocate the care of their owners: whether Andean herders might indeed consider them to be the "possessors of possessions". She suggests that Strathern's (1984) comments about the position of women in Melanesian society, as wealth, but not objects as postulated by our own subject-object matrix might provide a useful framework for thinking about animals in the Andes where they are both "part of and detachable from" human society.

Power, Politics and Balance

That science is posited as knowledge which is omnicompetent gives its proponents a unique authority to impose scientific analyses and solutions on the lives of people (and animals) in different parts of the world. As Penny Harvey has noted, the power of such expert knowledge "lies also in the entrenchment of social asymmetries between those who can impose singular truths and those who are subjected to the social consequences of these ways of knowing" (Harvey 2007:164). It is not unexpected therefore that in some form or other issues of power and politics are present in most of the chapters in this volume. Indeed, science and asymmetries of power, either between humans and animals or between different groups of humans, have already been touched upon in our discussions of the contributions by Dransart, Stephenson, Nimmo, and Bolton, while Degnen and Campbell examine the social and political debates that surround scientific research, animals and innovation.

As Harvey (2007: 180) also notes, abstract (rationalised) scientific knowledge is not intrinsically more mobile than other knowledges, for it still requires convincing social dramas to achieve wide reaching effects. Cristina Grasseni in her chapter looks at some of the social dramas surrounding technification and scientific innovation in the dairy industry in the highland areas of Lombardy in northern Italy and the social consequences of the changes they have brought about. The situation in Val Brembana is far from straight forward, with scientific solutions not so much being imposed upon local farmers as the remote, and seemingly traditional, farms themselves becoming sites of technological innovation. Grasseni documents how the boundaries of what constitutes the laboratory have become porous, as novel techniques are implemented on mountain farms and, faced with increasing economic and political pressure to increase productivity, highland breeders actively seek relations with biotechnology research institutions. Here animals and science stand for complex constellations of actors situated amidst considerations of technology, politics, economics and ecology, which Grasseni demonstrates through attention to the routines and relationships she observed and in which she participated as an ethnographer.

In drawing attention to the social ramifications of biotechnical innovations in farming, Grasseni notes that while the direction of scientific research is often highly controversial, and the subject of media attention and public scrutiny, questions about how and when to implement technoscientific innovation remain largely undebated in the public sphere. However, this does not mean that they have a negligible impact on the lives of people and animals. Although some of the farmers with whom Grasseni worked are eager to apply technoscientific innovation, the introduction of such measures has had a devastating impact on the farming community, with many smaller farmers having already been forced out of business.

While the exercise of political power *per se* is not so much the concern of Andrew Whitehouse, his contribution links to that of Grasseni through its exposition of a Latourian entanglement of science with politics and its discussion of the application of scientific measures in scenarios that affect livelihoods in rural areas. Whitehouse addresses the discourses and practices surrounding conservation strategies as ecologists, conservationists and gamekeepers debate what constitutes appropriate environmental relations on Scottish estates. Their particular concern is the management of populations of raptors and gamebirds and the interpretation of field studies into the effect of populations of birds of prey and of other factors on numbers of gamebirds present. Scientific data is mobilised and attributed different political significance by the different parties to the debate.

John Stuart Mill once remarked, in critiquing the ideas of his day on the need for balance in political constitution, that, "there seems to be something singularly captivating in the word balance, as if because anything is called balance, it must for that reason be necessarily good" (Mill 1988: 263). Whitehouse shows us how the idea of balance is mobilised by different actors in the raptor debates. Balance is considered something that is desirable and that humans should seek to foster in nature through their actions; science is mobilised by both conservationists and gamekeepers as a means of understanding how to generate a balance between populations of prey and predators that have been disrupted by what they deem to be the inappropriate actions of other humans. Balance becomes indexical of human behaviour, an indication of good land management. The debate is concerned not only about how best to achieve a balance of wildlife populations but also with what means are appropriate to balance the interests of estate managers, conservationists, raptors and gamebirds.

Balance reappears in other contributions in relation to science: science is seen sometimes as a means through which a balance with nature–or even a better-than-natural balance–can be achieved and at others its application is seen as an imposition that acts to disrupt a balance that should occur naturally. This is clearly true in the case of debates surrounding genetic modification, particularly in their treatment by Campbell. Classical science is indeed itself deeply permeated with the idea of balance: the notion that systems whether biological or physical will remain in or return to a state of balance or equilibrium unless acted upon by some outside influence (as in Newton's first law of motion, for example). This idea–that natural systems with no interference from humans will remain in a state of balance–appears to have become entrenched in some popular understandings of science.[5] Testament to this perception of balance as a naturally occurring state of affairs is provided both by the uptake among certain sectors of the public of James Lovelock's Gaia hypothesis of global homeostasis (Lovelock 2000), and also by those, often of very different political persuasions, who stubbornly remain unconvinced that human actions are responsible for climate change.

Ideas of balance in non-Western cultures also surface in the volume. The importance of achieving a balance in nutrition, or well-being, is an important focus of Bunn's discussion of Kyrgyz animal knowledge, while Bolton writes of the importance to Andeans of achieving balanced relations with supernatural actors and of this balance being disrupted by the introduction of modern methods of animal management.[6] There is a contrast to be made between these ethnographic studies of indigenous people and the Western examples. However, rather than drawing a naïve distinction between the West and the rest, it might be better to distinguish on the basis of these contributions between groups of humans who envision themselves as existing within a web of dependencies alongside other entities–which might include animal species, the earth, or supernatural beings–between which a balance should be sought, and those who consider themselves to exist outwith such a system. The latter category encompasses both those who wish to leave nature to its own course and those who see natural systems as being in need of manipulation and control in order to balance the interests of different groups of humans (the gamekeepers in Whitehouse's account, for example). Such a distinction, of course, returns us to the arguments of both Latour (2004) and Haraway (2008) and the different but related ways in which they seek to reconfigure human politics.

Human-Animal Relations

In an anthropological collection concerning animals and science, it would be surprising if animal-human relations did not feature strongly, as indeed they do. In the preceding discussion we have already mentioned many instances of the treatment of such relations in the chapters by Bunn, Dransart, Campbell, and others.

While several contributors contrast human-animal relations in Western cultures with those in different ethnographic settings, a somewhat different take on the question is provided by Cathrine Degnen's chapter. Degnen examines the absence of animals from British debates on genetic modification: while the British media has been eager to report on transgenic crops, such as genetically modified varieties of soya, maize and oil seed rape, there has been an absence of debate around issues related to animals. In particular, there has been almost no debate about "pharming", the use of transgenic (farm) animals to produce pharmaceuticals for the treatment of human ailments–a practice likely to become increasingly prominent in the near future.[7] This issue leads her to ponder the unusual place of farm animals in the imagination of the British public.

Degnen considers the implications of pharming for boundaries between humans and animals. Pharming posits animals as different from humans and so available for exploitation by them, yet through their similarity to humans–to the extent that they are able to produce enzymes and proteins normally specific to human bodies–they become imagined, in Degnen's words, as "quasi-human". There are further complications to the boundary in the case of farm animals. Cows, sheep, pigs and other farm animals already occupy an ambiguous position in the public imagination. They are an economic resource to be exploited, are rarely seen by many city-dwelling members of the British public, yet appear time and time again in children's stories and games where they provide lessons about human morals and values–a feature that George Orwell cleverly exploited when he came to write his own fable for adults, *Animal Farm* (Orwell 1973). Degnen asks how the perceived similarities of animals to humans together with the transgression of the animal/human boundary brought about by genetic engineering and the unique position that farm animals occupy in our culture have a bearing on public perceptions of the behaviour of the scientists responsible for the production of transgenic animals when they are seen as having transgressed the boundaries of acceptable human behaviour.

Milk and Mammals

Finally, and rather unexpectedly, milk, mammae and lactation emerged as a theme across a number of chapters in this volume. Perhaps in hindsight this recurring presence should not have come as such a surprise since milk is a vital substance that is life giving and it is also a substance produced by both human and non-human animals. For example, Bunn documents in her chapter the extensive Kyrgyz culinary practices related to processing milk and the products that are made from it for human consumption, milk that over the past 200 hundred years has increasingly come from cows and declined from camels and yaks, with mares' milk remaining in more constant use over the same time period. Milk products in this setting are an essential source of human nutrition (providing protein, vitamins, calories, antibiotics and fatty acids), but are also a critical component of a series of practices aimed at enhancing human well-being and restoring health.

Milk is however also a substance that calls attention to the shared mammalian grounding of human animals and non-human animals, at least within a Linnaean classificatory system. Milk thereby carries a heavy symbolic weight: think of Victor Turner and the Nkang'a puberty rite amongst Ndembu where the white milky sap of the Mudyi sapling plays a crucial role, symbolising amongst other things breast milk, nourishment and the link between mother and child (Turner 1967). It is thus a node of interconnection between species not just in regards to food, but also both in terms of fears of contamination (as Nimmo demonstrates in regards to zoonoses)[8] and as a potential source of medicinal therapies (such as the pharming of GM animals to produce proteins in their milk for human consumption as elaborated upon by Degnen).

While the symbolic links of milk might flow reciprocally between non-human and human, and back again, the traffic of both usefulness of milk products and fears over contamination move resoundingly in one direction only: for the benefit and comfort of humans. Also of note is the gendered nature of this relationship. The usefulness of milk brings *female* reproduction to the fore as without this reproductive force, milk would not exist. However, it is the processes owned (mainly by male scientists) that come to have value. As Dransart reminds us, the category "mammal" is itself a Linnaean classificatory term which is gender-laden as only the female members of the class are able to lactate, but it is a term that is used to encompass the entire class. She in turn links this gender-charged language to possession, naming practices of camelids by British owners, and local understandings of physiological knowledge.

Concluding Comments

We have come a long way from the existential angst of Sirius, Stapledon's dog with human capabilities created by science and positioned uneasily astride the human-animal boundary. The historical and ethnographic studies contained in this volume help illuminate how science's intersection with animals in Western settings has been instrumental in purifying that very boundary–by separating nature from human culture–yet how in the present, with new advances in biotechnology, that same intersection has destabilised and confused the separations we make between ourselves and other living beings. Furthermore, the contributions here examine how science, having posited animals as objects for humans to know, own and dominate, has also prescribed certain sorts of animal-human relations. This in turn has often led to attempts to reconfigure the relations of non-Western people with the animals amongst which they live. Through focusing on science, our contributors not only demonstrate that people elsewhere have different relations with, and knowledge of, beasts (and that different possibilities of relating to animals exist within our own Western worldview), but further suggest that our Western knowledge about animals and their positions in society, arrived at through Western science and the social sciences, is itself in need of rethinking to incorporate other ways of knowing. It is thus our contention in this volume that accounts in which animals meet science are starting to provide important theoretical insights for anthropologists and are setting new agendas for theory in anthropology and science studies. We hope you will agree with us on reading the chapters that bringing together what at first sight may have seemed an unlikely combination of anthropological concerns into one volume has been a worthwhile enterprise and that the richness of the themes evoked by the contributors to this volume attest to what can be achieved by putting non-human animals squarely in the frame of the anthropology of science.

Notes

[1] The original French edition of *Totemism* was published in 1962.
[2] http://www.socialsciences.manchester.ac.uk/pug/index.htm
[3] Epizootic**:** a temporary but widespread outbreak of a particular disease among animals. (Shorter Oxford Dictionary, Oxford: OUP).
[4] An early critique of humanism from this perspective is provided in Sartre's preface to Franz Fanon's *The Wretched of the Earth* (Sartre 1990: 22).
[5] With the recent development of chaos theory and theories of complex systems, which have been influential in both physical and biological sciences, science appears to have taken a course that departs from ideas of balance and self-regulation. However, we would argue that the idea of natural balance remains important to popular understandings of science.
[6] Although it could be argued that this balance is achieved partly through excess: excessive behaviour and in particular the excessive consumption of food and alcohol is deemed appropriate and necessary during festivities.
[7] Contrast the absence of debate surrounding pharming in animals with the publicity attracted by Kazuo Ishiguro's (2005) novel *Never Let Me Go*, set in a dystopian future in which humans are cloned as a source of organs for transplants.
[8] Zoonose**:** a disease communicable from animals to humans. (Shorter Oxford Dictionary, Oxford: OUP).

Bibliography

Archetti, E. 1997 [1992] *Guinea-Pigs: Food, Symbol and Conflict of Knowledge in Ecuador*. Oxford: Berg.

Bolton, M. 2006 "Genetic Defects or Generative Prototypes: Competing Models for Livestock Improvement in Southern Bolivia". *Journal of the Royal Anthropological Institute.* 12 (3): 531-549.

Bulmer, R. 1967 "Why the Cassowary is not a Bird: A Problem of Zoological Taxonomy among the Karam of the New Guinea Highlands". *Man,* New Series. 2 (1): 5-25.

Budiansky, S. 1992 *The Covenant of the Wild: Why Animals Chose Domestication*. New York: William Morrow.

Callon, M. 1986a "The Sociology of an Actor-Network". In M. Callon, J. Law and A. Rip (eds.), *Mapping the Dynamics of Science and Technology*, London: Macmillan.

—. 1986b "Some Elements of a Sociology of Translation: Domestication of the Scallops and the Fishermen of St Brieuc Bay". In J. Law (ed.), *Power, Action and Belief: A New Sociology of Knowledge?* London: Routledge.: 196-223.

Campbell, B. 2005 "On 'Loving Your Water Buffalo More Than Your Own Mother': Relationships of Animals and Human Care in Nepal".

In J. Knight (ed.), *Animals in Person: Cultural Perspectives on Human-Animal Interactions*. Oxford: Berg: 79-100.

Candea, M. 2010 "'I fell in love with Carlos the meerkat': Engagement and Detachment in Human-Animal Relations". *American Ethnologist* 37(2), 241-258.

Cassidy, R. 2002 *The Sport of Kings: Kinship, Class and Thoroughbred Breeding in Newmarket*. Cambridge: Cambridge University Press.

Cassidy, R. and M. Mullin (eds.) 2007 *Where the Wild Things are Now: Domestication Reconsidered*. Oxford: Berg.

Clutton-Brock, J. (ed.) 1989 *The Walking Larder: Patterns of Domestication, Pastoralism and Predation*. One World Archaeology Series. London: Unwin Hyman.

Descola, P. 2005 *Par- delà nature et culture*. Paris: Gallimard.

Descola, P. and G. Pálsson (eds.) 1996 *Nature and Society: Anthropological Perspectives*. London: Routledge.

Douglas, M. 1966 *Purity and Danger: An Analysis of Concepts of Pollution and Taboo*. London: Routledge.

Dransart, P. 2002 *Earth, Water, Fleece and Fabric: An Ethnography and Archaeology of Andean Camelid Herding*. London: Routledge.

Durkheim, E. 2001 *Elementary Forms of the Religious Life*. Oxford: Oxford University Press

Edelman, B. 2005 "From Trap to Lap: The Changing Sociogenic Identity of the Rat". In J. Knight (ed.), *Animals in Person: Cultural Perspectives on Human-Animal Interactions*. Oxford: Berg: 119-140.

Edwards, J., P. Harvey and P. Wade 2007 "Introduction: Epistemologies in Practice". In J. Edwards, P. Harvey and P. Wade, (eds.), *Anthropology and Science: Epistemologies in Practice*. Oxford: Berg: 1-18.

Evans-Pritchard, E. 1940. *The Nuer*. Oxford: Oxford University Press.

—. 1956 *Nuer Religion*. Oxford, Claredon Press.

Fisher, M.J. 2003 *Emergent Forms of Life and the Anthropological Voice*. Durham: Duke University Press.

Franklin, S. 2007 *Dolly Mixtures: The Remaking of Genealogy*, Durham: Duke University Press.

—. 2002 "The Anthropology of Science". In J. MacClancy (ed.), *Exotic No More: Anthropology on the Front Line*. Chicago: The University of Chicago Press:351-358.

—. 1995 "Science as Culture, Cultures of Science". *Annual Review of Anthropology*. 24:163-84.

Haraway, D. 2008 *When Species Meet*. Minneapolis: University of Minnesota Press.

—. 2003 *The Companion Species Manifesto: Dogs, People and Significant Otherness*. Chicago: Chicago University Press.

—. 1997 *Modest_witness@second_millenium.femaleman©_meets_Oncomouse™*. London: Routledge.

—. 1991 *Simians, Cyborgs and Women: The Reinvention of Nature*. New York: Routledge.

—. 1989a "Teddy Bear Patriarchy: Taxidermy in the Garden of Eden. New York City 1908-36". In *Primate Visions: Gender, Race and Nature in the World of Modern Science*. New York: Routledge: 26-58.

—.1989b *Primate Visions: Gender, Race and Nature in the World of Modern Science*. New York: Routledge.

Harvey, P. 2007. "Arresting Mobility or Locating Expertise: 'Globalisation' and the 'Knowledge Society'" In M. Lien and M. Melhuus (eds.), *Holding Worlds Together: Ethnographies of Knowing and Belonging*. Oxford: Berghahn:163-183.

Ingold, T. 1994 "From Trust to Domination: An Alternative History of Human-Animal Relations". In A. Manning and J. Serpell (eds.), *Animals and Human Society: Changing Perspectives*. London: Routledge. pp 1-22.

—. 1980 *Hunters, Pastoralists and Ranchers: Reindeer Economies and their Transformations*. Cambridge: Cambridge University Press.

Ishiguro, K. 2005 *Never Let Me Go*. London: Faber and Faber.

Keller, E.F. 1995 *Refiguring Life: Metaphors of Twentieth-Century Biology*. New York: Columbia University Press.

Knight, J. 2005a "Introduction". In J. Knight (ed.), *Animals in Person: Cultural Perspectives on Human-Animal Interactions*. Oxford: Berg.

—. (ed.) 2005b *Animals in Person: Cultural Perspectives on Human-Animal Interactions*. Oxford: Berg: 1-14.

Knorr Cetina, K. 1999 *Epistemic Culture: How the Sciences Make Knowledge*. Cambridge: Harvard University Press.

Kuhn, T. 1970 *The Structure of Scientific Revolutions*. 2nd edition. Chicago: University of Chicago Press.

Latour, B. 2005. *Reassembling the Social: An Introduction to Actor-Network Theory*. Oxford: Oxford University Press.

—. 2004 *The Politics of Nature: How to Bring the Sciences into Democracy*. Cambridge: Harvard University Press.

—. 1993 *We Have Never Been Modern*. Cambridge: Harvard University Press.

—. 1987 *Science in Action: How to Follow Scientists and Engineers through Society*. Cambridge: Harvard University Press.

Leach, E. 1964 "Anthropological Aspects of Language: Animal Categories and Verbal Abuse". In H. Eric (ed.), *New Directions in the Study of Language*. Cambridge: MIT Press: 23-63.

Lévi-Strauss, C. 1992 *The Raw and the Cooked: Introduction to a Science of Mythology*. Harmondsworth: Penguin Books.

—. 1963 *Totemism*. Translated by R. Needham. Boston: Beacon Press.

Locke, J. [1690] 1988. *Locke: Two Treatises of Government*. Cambridge: Cambridge University Press.

Locke, P. 2008 "Captive Elephant Management, The Tharu and The Nepali State". *IIAS Newsletter* 49:14-15.

—. In Preparation *Servants of Ganesh: Elephant Handlers in Nepal.*

Lovelock, J. 2000 *Gaia: A New Look at Life on Earth*. Oxford: Oxford University Press.

Midgley, M. 1994 "Beasts, Brutes and Monsters". In T. Ingold (ed.), *What is an Animal?* London, Routledge. 35-46.

Mill, John Stuart 1988 "Parliamentary Reform (1) 1824", *The Collected Works of John Stuart Mill, Volume XXIV, Journals and Debating Speeches Part 1*, Toronto: University of Toronto Press and London: Routledge and Kegan Paul.

Nader, L. 1996 "Introduction: Anthropological Inquiry into Boundaries, Power and Knowledge". In L. Nader (ed.), *Naked Science: Anthropological Inquiry into Boundaries, Power and Knowledge*. London: Routledge:1-25.

O'Connor, T. 1997 "Working at Relationships: Another Look at Animal Domestication". *Antiquity* 71 (271): 149-156.

Ortner, S. 1974 "Is Female to Male as Nature is to Culture?" In M. Z. Rosaldo and L. Lamphere (eds.), *Woman, Culture, and Society*. Stanford: Stanford University Press: 68-87.

Orwell, G. 1973 *Animal Farm, A Fairy Story*. Penguin Modern Classics. Harmondsworth: Penguin Books.

Philo, C. And C. Wilbert (eds.) 2000 *Animal Spaces, Beastly Places: New Geographies of Human-Animal Relations*. London: Routledge.

Pratt, M. L. 1992 *Imperial Eyes: Travel Writing and Transculturation*. London: Routledge.

Rabinow, P. 1999 "Artificiality and Enlightenment: From Sociobiology to Biosociality". In M. Biagioli (ed.) *The Science Studies Reader*: New York and London: Routledge. pp 407–416.

Radcliffe-Brown, A. 1952 *Structure and Function in Primitive Society*. London: Cohen and West.

Ritvo, H. 1998 *The Platypus and the Mermaid and other Figments of the Classifying Imagination*. Cambridge: Harvard University Press.

—. 1995 "Possessing Mother Nature: Genetic Capital in 18th-Century Britain". In S. Staves and J. Brewer (eds.), *Early Modern Conceptions of Property*. London: Routledge: 413-426.—. 1984 *The Animal Estate: The English and Other Creatures in the Victorian Age*. Cambridge: Harvard University Press.

Russell, N. 2010 "Navigating the Human-Animal Boundary". *Reviews in Anthropology* 39(1): 3-24.

Sartre, J.-P. 1990 "Preface". In F. Fanon, *The Wretched of the Earth*.Penguin Twentieth-Century Classics. Harmondsworth, Penguin Books: 7-26.

Sawday, J. 1995 *The Body Emblazoned: Dissection and the Human Body in Renaissance Culture*. London and New York: Routledge.

Shapin, S. and S. Schaffer. 1985 *Leviathan and the Air Pump: Hobbes, Boyle and the Experimental Life*. Princeton: Princeton University Press.

Shepherd, C. 2006 "From *in vitro* to *in situ*: On the Precarious Extension of Agricultural Science in the Indigenous World". *Social Studies of Science*. 36 (3): 399-426.

Smuts, B. 1999 "Reflections". In J.M. Coetzee, *The Lives of Animals*. Princeton: Princeton University Press: 107-120.

Stapledon, O. 1944 *Sirius: A Fantasy of Love and Discord*. Harmondsworth: Penguin Books.

Strathern, M. 1992 *After Nature: English Kinship in the Late Twentieth Century*. Cambridge: Cambridge University Press.

—. 1988 *The Gender of the Gift: Problems with Women and Problems with Society in Melanesia*. Berkeley: University of California Press.

—. 1984. "Subject or Object? Women and the Circulation of Valuables in Highland New Guinea". In R. Hirschon (ed.), *Women and Problems with Society in Melanesia*. Berkeley and London: University of California Press: 158-175..

—. 1980 "No Nature, No Culture: the Hagen Case". In M. Strathern and C. MacCormack (eds.), *Nature, Culture and Gender*. Cambridge: Cambridge University Press:.174-222.

Strathern, M. And C. MacCormack (eds.) 1980 *Nature, Culture and Gender*. Cambridge: Cambridge University Press.

Traweek, S. 1996 "Kokusaika, Gaiatsu, and Bachigai: Japanese Physicists' Strategies for Moving into the International Political Economy of Science". In L. Nader (ed.), *Naked Science: Anthropological Inquiry into Boundaries, Power and Knowledge*. London: Routledge: 174-197.

Turner, V. 1967 *The Forest of Symbols: Aspects of Ndembu Ritual*. Ithaca: Cornell University Press.

Verran, H. 2002 "A Postcolonial Moment in Science Studies: Alternative Regimes of Firing of Environmental Scientists and Aboriginal Landowners". *Social Studies of Science*. 32 (5-6): 729-762.

Viveiros de Castro, E. 1998 "Cosmological Deixis and Amerindian Perspectivism". *Journal of the Royal Anthropological Institute*. 4(3): 469-488.

Willerslev, R. 2007 *Soul Hunters: Hunting, Animism and Personhood among the Siberian Yukaghirs*. Berkeley: University of California Press.

CHAPTER ONE

COLONIAL INTERSECTIONS: THE DISSECTION OF ANDEAN CAMELIDS AND KNOWLEDGE PRODUCTION

MARCIA STEPHENSON

The preface to the first edition of one of the basic reference tools of veterinary medicine, Dyce, Sack and Wensing's *Textbook of Veterinary Anatomy*, includes as its epigraph a quotation from Goethe: "What one does not understand one does not possess" (Dyce *et al.* 2002: vii).[1] A statement such as Goethe's could well have served as a kind of catalyst to the professors and demonstrators of anatomy who worked at the Royal Veterinary College newly established in 1766 at the Château d'Alfort located on the outskirts of Paris. These were, of course, the years when the study of comparative anatomy was beginning to flourish, and anatomy and dissection were central practices for the acquisition of knowledge through direct observation and hands-on training. The art of dissection, moreover, required the cultivation of the different senses, emphasizing the importance not only of sight, but also of touch, smell, and even taste.[2] Barbara Stafford described how this sensorial practice helped shape eighteenth-century methods of inquiry:

> "The Galenic conception of anatomy as an 'opening up in order to see deeper or hidden parts' drives to the heart of a master problem for the Enlightenment. How does one attain the interior of things? Anatomy and its inseparable practice of dissection were the eighteenth-century paradigms for any forced, artful, contrived, and violent study of depths" (Stafford 1991: 47).

Stafford's observations call our attention to the violence surrounding the acquisition of knowledge, an association that is central to my essay. In this vein, Stafford goes so far as to describe the eighteenth century as the "second 'age of discovery'" due to its being the "'era of uncovering'"

(*Ibid*: 24). Dissection, thus, in all of its material and metaphorical implications of discovering, cutting, dividing, separating and analyzing, was intimately linked to the acquisition of knowledge in areas ranging from surgery to archaeology, and religion to philosophy (*Ibid*: 47).

And yet, this epigraph by Goethe placed at the beginning of a textbook on veterinary anatomy shows clearly another layer to the series of relationships among dissection and the violence subtending knowledge acquisition. Notably, it underscores how anatomy and dissection became paradigms for the prevailing notion of knowledge *as gain*, whether gain meant the acquisition of wisdom, experience, or wealth.[3] This association was particularly salient during the eighteenth century when scientific knowledge acquired through the systematizing of nature constituted a global project, as Mary Louise Pratt (1992: 29) has observed. The study of natural history moved beyond the coastlines of countries and continents to map the interior lands, their plants, animals and minerals (*Ibid*: 30). Understanding how dissection and the study of anatomy structured this form of knowledge acquisition enables a more nuanced reading of Stafford's comparison of dissection's "violent study of depths" to a "second age of discovery" and Jonathan Sawday's description likening the body in Renaissance culture to an unexplored territory, "a location which demanded from its explorers skills which seemed analogous to those displayed by the heroic voyagers across the terrestrial globe" (*Ibid*: 23). The epigraph by Goethe, in other words, helps to clarify how, during the eighteenth century, anatomy and dissection carried metaphorical, professional, and political implications underlying Western expansion (cf. MacLeod 1988: 2-4).

In the Andean region of South America, dissection–the violent opening up of the interior–serves as an appropriate metaphor for the changing dynamic of colonial relations. During the late eighteenth century, increasing numbers of foreign travellers, naturalists, and collectors made their way for the first time into the heart of the Peruvian viceroyalty. This new phase of exploration was central to what Pratt described as the intersecting processes marking both the appearance of natural history "as a structure of knowledge," and the shift in attention away from maritime exploration and toward the interior highlands of the cordillera and puna regions. This distinct movement away from the coastal areas and towards the interior highlands formed part of a "new territorial phase of capitalism," one that was set into motion by the desire to acquire raw materials and useful objects of natural history. These foreign voyagers were impelled forward by "national imperatives to seize overseas territory in order to prevent its being seized by rival European powers" (Pratt 1992: 9).

Among the many objects of natural history that caught the interest of French *voyageurs naturalistes* and other entrepreneurs were the Andean camelids, including the domesticated llamas and alpacas and the wild vicuñas and guanacos. French travellers perceived these animals to be a promising source of unexploited riches. As early as 1741, the *Dictionnaire Universel de Commerce* (1741: 624) listed entries on the camelids, calling particular attention to the economic importance of the fibre of the vicuña. In 1779, the Abbé Beliardy argued in favour of France acquiring three or four dozen female vicuñas and half a dozen males in addition to a number of alpacas and llamas for the purpose of acclimatization, claiming that "the Minister who would contribute to enriching the kingdom with such a useful animal [as the vicuña] could congratulate himself just like for the most important conquest."[4] Beliardy's remarks could well have gone unnoticed had the comte de Buffon not quoted them extensively in his observations on the vicuña, which he published in 1782. Buffon cited Beliardy at length because both men believed that for the advancement of science and public welfare it would be an important project to bring llamas, alpacas and vicuñas to France to naturalize them so that their valuable fibre could be more easily exploited. Consequently, Buffon's study of the natural history of the vicuña was also written to convince the government to support this economic project (Buffon 1782: 210-211). Beliardy's simile likening the acquisition of the animal to a conquest underscored this confluence of colonial and scientific enterprises.

Donna Haraway's work on primates helps us to understand how the Andean camelids played a crucial role in "the system of unequal exchange of extractive colonialism," and their bodies, living and dead, figured prominently in discourses that depended "on a flow of value from the lands" where the animals lived "to the lands where they were exhibited and textualized" (Haraway 1989: 19). Dissection paradigmatically structured the narratives of acquisition of the Andean camelids during the eighteenth century because it entailed a series of "'surgical strategies'" for the purposes of cutting, extracting, and even transplanting.[5] At the same time, dissection was also closely related to the new technologies of preservation whereby the surviving remains of the anatomized body became an archive for the consolidation of memory, power, and knowledge. This chapter examines the relationship between dissection and colonialism by examining three specific instances where the anatomization of a camelid body took place. It argues that dissection and preservation became key paradigms during the eighteenth century for understanding how camelids and their bodies conveyed meaning in the violent production and acquisition of knowledge for Europeans as well as for indigenous peoples.

Dissection in the Contact Zone[6]

One of the first French travellers to explore the inner coastal regions of South America was the priest and mathematician-astronomer Louis Feuillée (1660-1732). Sent by Louis XIV, Feuillée spent about eight months in the area in 1709-1710, recording his observations on a variety of topics related to cartography and the natural history of the region. During the course of his travels he had the unexpected opportunity to dissect the digestive tract of a guanaco. The animal had been given as a gift to his captain but it died unexpectedly shortly after. For European naturalists and anatomists, Feuillée's is the earliest known written description of the dissection of an Andean camelid.[7] Feuillée's work was particularly significant because of the epistemological transformation that it set into motion. In his account, dissection can be read as a metaphor of anatomization and extraction whereby the priest slowly excised the animal from its natural environment and cultural role in indigenous ritual practices in order to transplant it into a new context of European colonial expansion.

Feuillée first encountered the camelids when he noticed what must have been a caravan of llamas accompanied by two or three Indians, approaching from the distance: "I judged by the route they were taking that they would pass close to my tent. I returned there quickly, desiring to see them up close so I could examine them. I knew from the report I had been given that their face was quite extraordinary and I wanted to do a sketch" (Feuillée 1725: 21).[8] While observing that the drawing of the llama made by his adversary, Amedée Frézier, was true to form (*Ibid.*), Feuillée added further details in his journal regarding the animals.[9] Basing his account on early colonial sources such as the Inca Garcilaso de la Vega and Blas Valera, the French priest described the natural history of the camelids and their economic and cultural significance for Andean indigenous peoples, lingering on the close relationship between humans and animals, as well as on the role the camelids played in the foundational myths of origin and native ritual (*Ibid*: 22-24). Feuillée concluded his remarks for 31 August 1710 noting that Frezier's description of these Peruvian "sheep" had been written with enough truth that there was no need to spend time on further reflections (*Ibid*: 24)

The following day, September 1, Feuillée opened his journal with a brief reference to the weather and then he recounted the story of the gift of the two guanacos, a male and a female, that his captain received and which he planned to take back to France (*Ibid*: 25).[10] One of the animals died that morning, so Feuillée took advantage of these unfortunate circumstances and immediately set about dissecting the digestive tract of the dead animal. He worked quickly because of the fast onset of putrefaction, and, even so, his plans to finish the dissection on the following day had to be cancelled due to the horrible stench of the body which prevented anyone from approaching it (*Ibid*: 32).

Interestingly, the original motive for cutting open the animal was not science, but commerce: Feuillée was looking for bezoar stones (*Ibid*: 25). Europeans believed that the highly prized bezoars, or calcinated concretions found in the camelid gut, constituted a powerful antidote for poisonings of different kinds and a valued remedy for a variety of illnesses ranging from melancholy to typhoid fever. Due to their rarity, they continued to bring large sums of money in the Americas and Europe, even as late as the eighteenth century.[11] When Feuillée opened up the first compartment of the guanaco's stomach, however, he was disappointed because instead of the treasured bezoar stone what he found was a large quantity of poorly digested food (*Ibid*: 28). The end result of an otherwise fairly unpleasant experience constituted the first and last written account of the animal's digestive tract from the eighteenth century.[12] His extensive description of the compartments of the animal's stomach and the mechanistic, engine-like manner of digestion was not superseded until the mid-nineteenth century. The celebrated French comparative anatomist Georges Cuvier had access to an infant llama while preparing his *Leçons d'Anatomie Comparée* (Cuvier 1835: 72-75), but his comments on the camelids were considered to be incomplete, and so Feuillée's description remained the primary model to which later anatomists would continue to refer.[13]

Feuillée's text is significant not only because it described in detail the dissection of a camelid. Just as important, the account of the dissection produced a subtle yet decisive shift in focus with regard to the place of the animal in the priest's narrative. Although at the beginning of his writings on the camelids, he depicted the animals in the context of native ritual practices and foundational narratives, during the course of the dissection, the presence of native peoples was carefully excised and replaced by the language of European science.[14] Pratt has observed how "the conventions of travel and exploration writing (production and reception) constitute the European subject as a self-sufficient, monadic source of knowledge" (Pratt 1992: 136). Any interactions that the traveller may have had with indigenous peoples or other local inhabitants disappear, only to emerge here and there in the text as mere traces (*Ibid*: 136). The presence of indigenous peoples at the scene of dissection in Feuillée's text was briefly alluded to when the French priest mentioned in passing that it was actually an Indian who showed him which compartment of the stomach contained bezoar stones.[15] The Indian's presence in the text as a guide became unnecessary once he indicated where the stones are generally found. Because his knowledge of the animal's anatomical profile was no longer of consequence to the narrative, he soon vanished from it altogether. Feuilleé's account of the dissection of the guanaco went even further when it discursively severed the animal from its native habitat only to resituate it at the centre of European scientific and commercial enterprises. His detailed mapping of the animal's internal geography would subsequently assist comparative anatomists in Europe as well as foreign travellers and hunters in search of bezoar stones.[16] In this context, Stafford's observation that dissection can be a metaphor for "an intellectual 'taking to pieces'" (Stafford 1984: 80) suggests that in the context of Western colonialism, it can also be a metaphor for the processes whereby Western scientific and commercial practices take to pieces Andean traditions and forms of knowledge. Following Feuillée's account, the animals gained in importance to the foreign *voyageur-naturaliste* not so much for their role in native cultural and economic traditions but rather for their role in colonial economic expansion.[17]

The process of separation by means of anatomization was described in the guides and manuals appearing in the late eighteenth and early nineteenth centuries. These pocket-sized books provided detailed instructions to travellers on how to collect and preserve natural specimens. Prior to the mid-nineteenth century, explorers and naturalists travelling through the Andes had a difficult time transporting live animals. Consequently, it was important to know which parts of an animal and what kind of information should be collected in the field to take back to the natural history collections and menageries of Europe. In his "Directions for Preserving Animals, and Parts of Animals for Anatomical Investigation" (Hunter 1809), the acclaimed Scottish anatomist John Hunter (1728-1793) gave precise instructions to "Travellers of Research and public spirit" on how to prepare for transport mammals that the collector hunted or trapped. He noted how "At the time of taking an Animal, it would be proper to collect, on the spot, as many circumstances of its History as possible, relative to Food, Propagation, Locomotion, etc. &c" (Hunter 1809: 7). Even though the animal was dead, much valuable information could be gathered that would eventually assist efforts to transport the live animal and ultimately acclimatize it in a new land. Hunter recommended that "When an Animal is opened, for the Purpose of Separation, it will be proper to take a general

View of the Parts in their natural Situation...Animals, whose Food is not exactly known, should have the Contents of the Stomach examined, to learn, if possible, what Aliment they had taken last; and also of the Colon and Rectum, to determine the Kind of Feces they contain" (*Ibid*: 12; see also Capus 1883; Turgot 1758; and Boitard 1835). Hunter's instructions to the "traveller of research and public spirit" were fairly typical insofar as they too left out of consideration native or local guides and informants from the collection scene. Instead, Hunter instructed the traveller to rummage through the digestive tract and faecal material as if knowledge of the animal were self-contained in its dead body. Clearly, it was hoped that careful dissection and close examination would be sufficient to bring this information to light.

Just as Hunter's instructions directed the traveller's method of dissection, they also highlighted the importance of collecting and preserving animal fragments. Writing more generally on animal collection in Europe and the United States, historian Nigel Rothfels called attention to the significant role animal fragments play because of the narratives they produce:

> I've thought about the literally millions of animal remains in American and European museums and realized that almost every one of them must have a small tag upon which one could read where, when, and by whom the animal was collected. Every time I have begun to consider these animal fragments, I've come to the conclusion that these animals' *un*natural histories–their lives and afterlives not in their 'native haunts' but in such human environments as museums, books, circuses, and zoos–are at least as worth telling as their natural histories. (Rothfels 2002: 6).

Rothfel's use of the terms "animal fragments" and "animal remains" underscored the importance of anatomization and preservation and the full range of their literal and metaphorical meanings for the narratives that they generate. In the section that follows, the essay will take up the second example from the eighteenth century where a camelid was artfully dissected and, also, carefully preserved: the case of the "Fragonard llama." I contend that European aesthetic systems of the time–notably picturesque anatomy, drawing, and travel - produced narratives based on these animal remains that helped to make the llama more fully "knowable" and hence useful to larger colonial projects of acclimatization and naturalization.

European Aesthetics of Dissection and Preservation

Although the Abbé Beliardy had called for the introduction of live specimens into France, the transportation of camelids from the Andes to the port cities of Europe posed a number of complications for collectors and merchants as well as for ships' captains charged with the animals' welfare on the return voyage across the Atlantic. As happened with Feuillée's guanaco, camelids that travellers herded through the mountains and down to the coasts with the intention of taking back to Europe frequently died while *en route* over land. The mortality rate was even higher on board ship because the majority of the animals carried in commercial and military vessels around the bottom of Cape Horn simply could not survive the long, gruelling journey.[18] Due to the fragility of the camelids and the difficulties involved in their transport, especially of the wild vicuñas, it is hardly surprising that few of the animals actually reached Europe alive.[19]

While living examples of the Andean camelids in Europe during the eighteenth century were few and far between, Georges Louis Leclerc, comte de Buffon, had access to both a llama and a vicuña while writing the Supplements to his monumental *Histoire naturelle, générale et particulière*. Buffon studied these two animals while they lived at the veterinary school at Alfort. At the time, the veterinary school maintained a menagerie of domestic and exotic animals so that students could work on acclimatization projects as well as on the clinical study of diseases that affected exotic species (Railliet and Moulé 1908: 378). According to Buffon's record, a young male llama and its mother were sent to Alfort from England in November 1773. The date of arrival of the vicuña is less clear. Buffon noted only that it was drawn while alive in 1774 and that it lived at the veterinary school for fourteen months and most likely had spent the same amount of time beforehand in England (Buffon 1782: 208-210). Unlike the young llama, which was kept almost like a pet, permitting its keeper and others to climb on its back (*Ibid*: 206-207), the vicuña was described as vicious and ungovernable, constantly trying to bite its care givers and repeatedly spitting all over them (*Ibid*: 210; Boitard 1842: 427). When Buffon was writing his account of the animals in August 1777, the male llama was still alive but the vicuña had since died and could be seen, stuffed, on display in the museum or anatomical cabinet of the school's founder, Claude Bourgelat (Buffon 1782: 208).

Of particular interest to this chapter is the fate of the mother of the male llama. She apparently survived the trip from England only to die shortly after arriving at the veterinary school. She, too, ended up in

Bourgelat's anatomical collection, but she was not stuffed. Buffon noted in his first paragraph on the young male llama that the mother had been skilfully dissected and preserved (Buffon 1782: 204), her muscles and organs treated, dried and varnished, the veins and arteries injected with blue and red resins.[20] She can still be found today at the Alfort veterinary school in what is now called the Fragonard Museum, in honour of the anatomist responsible for creating most of the original, preserved specimens. A closer examination of the "Fragonard llama" provides a striking example of how the discourses of science and colonialism converge at the site of dissection. In this case, the anatomist moved beyond Feuillée's discoveries of the digestive tract to uncover the animal's entire body by flaying it and subsequently exposing the vascular and musculoskeletal structure. Once the body interior was fully surveyed, the animal could be appropriated for its usefulness in future commercial enterprises and European acclimatization projects.[21]

Honoré Fragonard, cousin of the famous painter Jean-Honoré Fragonard, was the eccentric Director of the Alfort Veterinary School and the principal anatomist and demonstrator from 1766-1771, when Bourgelat dismissed him for reasons of insanity attributed to kidney stones.[22] Unlike other anatomists of the time who taught their students with the use of illustrations and engravings, and wax and plaster models, Fragonard preferred to work with the "original tissue and living skin" (Sournia 1981: 5) of the animal and human cadaver, perfecting the method of wax injection used at the time for the preservation of organs and different parts of the body.[23] Fragonard was not, of course, the only anatomist to use dissection and wax injection for preserving dry specimens. Indeed, these methods had been in use since the previous century. What made Fragonard's work so distinct was that in addition to preparing a specific organ or part of the body, he expertly preserved the entire body with his carefully guarded formulas (Simon 2002: 76). Most of the injected specimens were kept for instructional purposes, and students were generally involved in the preparation process. Bourgelat, a proponent of training students in anatomical preservation, observed in the 1771 *Réglement des Écoles*, that the best way of getting the students' attention was by having them clean bones. Those who showed a talent for this work could then receive further instruction (quoted in Degueurce 1999b: 93).[24]
While many of Fragonard's carefully dissected and preserved works were prepared for the purposes of teaching, other creations achieved startling effects because they were produced with flayed bodies deliberately arranged in theatrical positions. In his biographical remarks on the anatomist, Michel Ellenberger contends that even today the visitor who

first walks into the Fragonard museum experiences somewhat of a shock upon seeing such macabre yet artful spectacles. For example, the flayed llama looks as if its skin had been turned inside out and stretched (Ellenberger 1981: 8) (see Figure 1-1).

The presence of exotic animals such as the flayed and injected llama and stuffed vicuña and the theatrically displayed human anatomies, especially Fragonard's famous "horseman of the apocalypse" which included a complete flayed horse and rider (holding blue silk reins), attracted visitors from different parts of Europe who travelled to Alfort specifically to visit the collection. Although reports published at different times indicate that the cases of display were rearranged, the llama generally appears to have been placed in a fairly central and visible location, often near the horse and rider, one of the collection's big draws (Langlois and Lévesque 1804: 33). Chabert, Flandrin and Huzard described anatomical displays of exotic animals, including the llama, placed in opposition to the horse, with human anatomy commanding the attention of all (Chabert *et al.* 1782-90: 17). Christophe Degueurce, curator of the museum, has suggested that the Fragonard collection cannot be easily categorized. Positioned somewhere between the pedagogical laboratory and the cabinet of curiosities, it reflects changing perceptions in the understanding of veterinary medicine as well as in cultural reception of disease and anatomical display (Degueurce 1999b: 91-92). Simon similarly acknowledges the "multiplicity of functions that these objects perform, and hence the numerous spaces they occupy. It is only at the intersection of these multiple spaces (rather than at the more commonly cited crossroads of art and science) that we can begin to understand their apparently monstrous appearance" (Simon 2002: 69).[25]

Taking Simon's analysis one step further, I would argue that it is only by looking at these intersections and multiple spaces that we can see how colonialism is embedded in the aesthetics of art and science. The body of the llama, flayed, dissected and injected to reveal its internal geography, visibly expresses this confluence. By means of his accomplished techniques of preservation, Fragonard created a particular aesthetics that shared certain elements with the field of picturesque anatomy, in vogue at the time, and with the aesthetics of picturesque landscape painting, also coming into fashion during the late eighteenth century. It is in the aesthetics of the picturesque that science, art, and colonialism come together to appropriate the camelid body as a promising resource for exploitation in Europe.[26]

Figure 1-1: Head of the Fragonard llama (photograph: M. Stephenson)

Picturesque, or artistic, anatomy combined the study of dissected cadavers with living models to give the student of painting, sculpture, or medicine a better working knowledge of the details of the skin, muscles, and skeletal structure as well as of the changes effected when the body is in motion. The famous French anatomist, Jean-Joseph Sue (the elder; 1710-1792), in addition to being professor of anatomy at the Collège Royal de Chirurgie, was also professor of anatomy at the Academie Royale de Peinture et de Sculpture where he offered courses each year on picturesque anatomy. Picturesque anatomy emphasized close examination of the body's deep, internal structure, one that could only be visualized through dissection.[27]

John Hunter's equally famous brother, the anatomist William Hunter, taught courses in picturesque anatomy in London. He perceived the study of anatomy, like that of sculpture, wax modelling, and landscape painting, to be three dimensional (Jordanova 1985: 395). Ludmilla Jordanova's description of Hunter's work could also be applied to Fragonard's aesthetic project:

> "Hunter's enterprise was closely akin to that of a cartographer, lovingly recording all the details of terrain–flesh and tissues. There was no smoothing out here but, rather, corrugations depicted in loving detail. When the mapping had been completed, the human [or animal] body would, in some significant sense, be known and understood" (*Ibid*).

The addition of colour to wax injections was key to observing and tracing the contours of the body's geography. For example, Francesco Algarotti in his *Essai sur la Peinture, et sur l'Academie de France, établie à Rome* (1769) suggested that the use of different colours to distinguish muscle groups, and arteries from veins, would facilitate the process of memorization for the artist and medical student. Students could thus read the injected bodies much as if they were coloured maps that showed the delineations of provinces and states belonging to a particular prince (Algarotti 1769: 22-23). As these examples indicate, knowledge of the body through its mapping enabled it to be more easily claimed or possessed. To facilitate this process, John Hunter stressed the importance of selecting colours and shades that would be pleasing to the observer's eye when preparing to inject the vascular and arterial systems of the anatomical specimens: "To these injections add as much colour as makes them appear of a proper bright colour to the eye...It would seem that we have not yet any good green colour; but as blue and yellow make a green, blue verditer, added to yellow wax or resin, gives us a fine green" (Hunter 1861: 391). Commenting on the importance of colour, Philippe Ariès noted how the beautiful hues of a cadaver just beginning to decompose could achieve the subtlety of a Rubens or Poussin (quoted in Blin 1981: 6).[28]

Like artistic anatomy, picturesque painting emphasized rugged, irregular landscapes, the nuance of colour, and the subtle play of contrast between light and shadow (Gilpin 1792: 20). In her description of the picturesque, Stafford (1984) gives particular attention to its theatrical and exotic elements. In garden theory, for example, she points to the presence of

> "An engaging theatrical vocabulary of caves, rocky cliffs, sylvan settings, and inlets - all stressing a wild, untamed environment... It was crucial to the formulation of garden theory that the representation of entangled wilderness depended on the handling of nature in such a way as to produce dramatic, if not spectacular effects" (*Ibid*: 6).

Fragonard's aesthetics of preservation depended on the rough, irregular textures of tissue and muscle and the contrast of colours, and so his *écorchés* were also dramatic and spectacular. For example, when applying coloured wax injections, he deliberately enlarged the arterial and vascular systems so that they were much more swollen than normal, thereby creating startling images of flayed human and animal faces with almost painfully engorged arteries and veins.[29] Simon notes that the effect of theatricality was emphasized further with the way the rough and uneven sheets of dried and varnished muscle "reinforce the impression of a tortured body, flayed rather than dissected. In this sense, the pieces are reminiscent of many Christian images depicting the physical suffering of martyrs" (Simon 2002: 69-70). Fragonard's creations thus contrasted sharply with the "neatly pinned-back skin of anatomical drawings or wax models" (*Ibid*: 69). Instead, they express what Sournia has described as a "pathetic subjectivity" because their haggard faces reflect the imminence and anguish of death (Sournia 1981: 5). Or, they show the phantasmagorical movement of death coming back to life, as Blin has argued (*Ibid*: 6). Like a surgeon or sculptor, Fragonard masterfully brought dissected flesh to life (Ellenberger 1981: 10). Perhaps, most troubling, however, are these creatures' dissected, lid-less eyes that look out onto the world with a blind, fixed stare (*Ibid*: 8).

Fragonard's aesthetics of dissection and preservation thus emphasized both movement and stasis. The play of colour and texture of these extraordinary three-dimensional creations such as the llama, enticed the eye to travel across the picturesque landscape of the flayed body. So, too, when he examined Fragonard's art, Jean-Joseph Sue spoke of the "'concert' of movement of the fluids in the vessels" (cited in Sournia 1981: 5).[30] And yet, the dissected specimen was also frozen in space and time, on display as if it had been captured and immobilized by the glass enclosure of the museum cabinet (see Figure 1-2). Paul Carter's analysis of the aesthetic of the picturesque and Australian colonial history similarly indicates how, through the deployment of metaphors of mapping and enclosure, the picturesque participates in the possession and domestication of "wild" nature.[31] For example, both picturesque travel and gardening reflect an analogous dialectic between movement and rest, travel and settlement. In a colonial context, however, the dynamic takes on additional significance: "Implicit in both spatial modalities was always the sense of symbolic boundaries defined and rendered eloquent. The screen of vegetation, the trees one would not wish to see cut down, might, in other contexts, be a bar to physical and imaginative progress. To call them picturesque was to attribute to them the observer's own heightened sense of possession, his sensation of suddenly being at home in the world" (Carter 1988: 242-243). Drawing from Carter, it can be argued that picturesque aesthetics structured the processes enabling the colonized

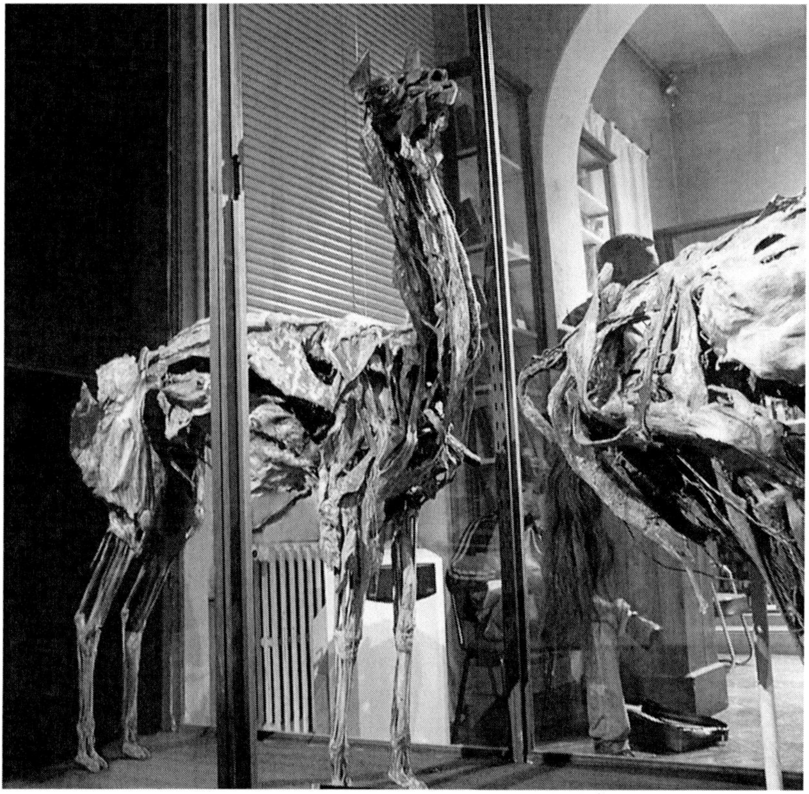

Figure 1-2: The Fragonard llama (photograph: M. Stephenson)

body-landscape to be explored, enclosed, *and* possessed. Recalling Goethe's epigraph once more, we can look at the llama in the glass cabinet as a kind of archive produced through the art and science of dissection and preservation.

Like Algarotti's colourful, mapped anatomies, the Fragonard llama becomes a visible record testifying to the aesthetic moment of colonial knowledge production-as-possession.[32]

Dissection, Preservation, and Knowledge Production in the Andes

The final section of this chapter turns to the Andean region to consider the depiction of the sacrifice of a camelid by an indigenous artist-chronicler,

Felipe Guaman Poma de Ayala (1980). In contrast to the two previous dissections analyzed, Guaman Poma's work is not from the eighteenth century. Yet, his illustration is unique because it is one of the only visual images produced of this ritual event prior to the twentieth century. Included in his manuscript-letter *Nueva corónica y buen gobierno*, written to the Spanish king Philip III in 1615, Guaman Poma's representation of the sacrifice of a llama by Andean indigenous peoples provides a third, compelling example of how knowledge is produced at the intersection of multiple spaces in a colonial context.

The llama, described by many as the animal most perfectly suited to the ecology of the Andes, has played a critical material and symbolic role in the political economy of Andean social and cultural practices. The animal's meat, fat, bones, and excrement were used by herding communities and they also figured importantly within a complex system of exchange networks developed throughout the Andes, the Pacific coastal region, and Amazonian basin. The camelid's fibre served as the basis for clothing, rope, and other practical items, and as a fundamental component of the complex weave of Andean memory practices. For example, the famous *kipus* or knotted cords that served as archival documents were made from camelid fibre (see Arnold and Yapita 1998: 18), and the animals themselves played a prominent role in ritual sacrifices that linked native Andeans to what Tom Abercrombie has described as their "meaning world" (Abercrombie 1998: 130). Andeans perceived that their lives were interlinked with the llamas and alpacas, which were believed to be on loan from the gods for as long as they took care of them and treated them with respect. In return, the camelids accompanied them along the way from their origin times and on into the next world (*Ibid*: 183; Flores Ochoa 1977: 227). This helps us to understand why "llamas were (and are) the *sine qua non* of sacrifice, in part because Andeans have envisioned llamas as enjoying a kind of social life comparable to that of humans" (Abercrombie 1998: 183).[33]

In his *Comentarios reales de los Incas* (1609), the Inca Garcilaso de la Vega described having witnessed clandestine sacrifices of camelids that were carried out by old Indians who had not yet been baptized (Garcilaso 1985 II: 51). What could be called native forms of dissection were forbidden by the Catholic Church because they were deemed to be idolatrous practices. In part, this was due to the fact that some ritual sacrifices began, not with the slitting of the animal's throat, a practice familiar to Europeans,

Figure 1-3: Guaman Poma's depiction of the sacrifice of a camelid (Guaman Poma 1980 II: 273 [880]. Reproduced with permission of The Royal Library, Copenhagen)

but instead with an incision at the end of the ribcage.[34] The officiant could then insert his hand and carefully extract the llama's heart, lungs, and oesophagus without making a second incision or cut of any kind. The veins, arteries and organs would be inspected to determine if they augured good or bad fortune for the community (*Ibid*: 50-51). For example, if the pastures had been ample and abundant and the animal was free of life-threatening parasites, the organs would be healthy and strong, indicating prosperity for the entire herd. Similarly, if the community's territorial extension had been encroached upon and pasture lands were insufficient or poor in quality, or if there had been unfavourable climatic conditions, the health of the herd and, consequently, of the community, would be adversely affected.[35] Following this inspection, the organs would generally be burned and the animal roasted and eaten. As this complex ritual indicates, sacrifices of camelids form an integral part of what Sawday has described in another context as a "culture of dissection," which consists of a "network of social practices, social structures, and rituals surrounding this production of fragmented bodies" (Sawday 1995: 2).[36]

The social network forming part of the Andean culture of dissection is visibly foregrounded in Guaman Poma's image (Guaman Poma 1980 vol. II: 273 [880][37]) (see Figure 1-3). He depicted a family unit, consisting of mother, father, and son, gathered together to sacrifice the animal. The son, positioned on the left, helps to hold the animal as the father, occupying the central focus of the image alongside the animal, carries out the actual anatomization; the mother stands close by on the right, holding a container and waiting for the internal organs. The image could be categorized as one of the depictions of Andean idolatrous practices, practices which the author generally condemned. At the top of the page he wrote: "They kill the ram here as in the time of idolatry, by inserting their hand near to the heart.[38] Let them not kill the animal this way, but rather as in the time of Christianity, by cutting the throat. It is idolatrous witchcraft to kill the animal in the old way. Let the Indian man and woman who carry out this practice be punished" (*Ibid*). On the following page, the textual exegesis of the image continues to warn against such methods: "The Indians of this kingdom are carrying out the old tradition of idolatry; because, either for eating or for rituals, they killed sheep, they opened it up by the heart, which is the tradition of idolatrous sorcerers" (*Ibid.*II: 274). Following this warning, the author includes a brief dialogue between the son and the father: "the youth said to the old man: ... 'they died by their heart, father'. ...We will be well off, son... we will eat blood-mix [bollos de sangre] and raw blood'; and so they first ate raw blood and flesh, later they ate cooked meat, according to the law of idolatry, and so, from now on, do not kill sheep by the heart, rather slit their throat and do not eat raw blood or meat, and let those [who disobey] throughout the kingdom be shorn and punished in the streets with fifty lashes" (*Ibid.*II: 274).

Guaman Poma's textual rebuke of the practice appears to be concerned primarily with the idolatrous *method* of killing rather than with the act of killing itself, and in the consumption of uncooked meat and blood. The visual image focuses attention on the method of anatomization and to the social networks created as a result. Its impact is powerful, eliciting, perhaps, a reaction of horror due to the apparent brutality of the vivisection, mixed with compassion for the animal. Just as we saw with the Fragonard llama, Guaman Poma's llama is captured in its final moments of agony, literally at the threshold between life and death: the wide eyes bulge, the gasping mouth exposes incisors and canine teeth. Teetering on tied legs, the animal defecates as the body relaxes its grasp on life and death takes hold. Here, too, the llama undergoes the physical torment of martyrdom, the bloody wound in his side a symbol of the passion of Christ.

How might Guaman Poma have wanted the king of Spain to understand this image? One approach to this question is to compare the author's rendition of the camelid dissection with the Roman sculpture of an old man flaying a young goat (see Figure 1-4).[39] The sculpture portrays the act of the animal's evisceration. A sole protagonist, the man is dressed in skins, the texture of which matches the wild hair and beard framing his leering face and rolling head. Showing no sign of movement, the animal hangs lifelessly from the tree where its hind legs are tied. The man's left hand and arm hold the animal in a tight embrace, while the right hand plunges inside the abdomen, tearing out its entrails. Smooth and muscled surfaces alternate across the man's bare chest, arm and leg, echoing the contours of the flayed animal's body. The animal's visible genitalia suggests a prior contest of strength between the two, but, having lost, its face is now hidden to the viewer by the flaps of loose skin. Although the man's genitalia are covered, the viewer perceives a certain kinship between the two due to their intimate closeness and the insinuation of a shared animality.

Guaman Poma was certainly familiar with Western conventions depicting savagery and barbarism, and he was a master at turning negative images of Amerindians on their head, as Rolena Adorno has shown (1990). For example, his portrayals of pre-Inca Andean peoples depict them wearing elaborate skins but carrying out civilizing activities such as ploughing or Christian-like activities such as praying (*Ibid.*). The author's artistic rendition of the camelid sacrifice is similarly complicated because the visual signs generate multiple possible readings.

Figure 1-4: Roman sculpture of an old man flaying a goat (Louvre Museum, Paris – photograph: M. Stephenson)

Adorno (1979) has described in detail Guaman Poma's elaborate pictorial system of iconographic representation and spatial organization. Of particular interest to this essay is how he positions images of positive value (Christian faith, spiritual knowledge, authority, etc.) on the left hand side of the page (on the right hand side from the point of view of the image) versus images of negative value (corruption, disorder, depictions of hell, cruelty, transgressive behaviour, etc.) on the right hand side (the image's left hand side) (see Adorno 1979: 33; 37-43). Many elements of the image of the llama sacrifice corroborate this spatial politics. For instance, the old man who is carrying out the idolatrous action has a leftward, negative, orientation (looking toward the viewer's right). The positioning of his hand subtly reinforces the transgressive nature of the act. Buried inside the camelid's body, it also points leftward toward the intestines rather than toward the more "noble" organs of the heart and lungs. Yet, in this case, the man's face shows neither pleasure nor erotic enjoyment of the act of violence. Rather, its concentrated expression is more suggestive of the detachment of a surgeon making the first incursion into another's body. He skilfully navigates his way through the animal's internal geography by feel rather than by sight, a sign of both experience and expertise. The hand hidden inside the animal's body does not provide so much as a glimpse of what he may be about to pull out, thereby mitigating the impact of the image. The family gathers closely around the body of the struggling animal, each member playing a specific role in the anatomical procedure. In spite of his stern condemnation of the method of anatomization, Guaman Poma nevertheless depicts an intimate scene that underscores not so much the family's shared idolatry, but rather the vital social relationship between animals and humans.

It may be difficult to perceive this important relationship on the basis of this image alone. Looking at the entire set of images from Guaman Poma's 1615 manuscript, however, one begins to understand the close relationship between humans and camelids. As depicted in image [225], young girls of the age of twelve or so, served their families by watching over the herds of camelids while spinning; boys did the same, while learning how to hunt for birds [204]. Camelids sang in chorus with humans during ritual celebrations [318] and wept with them as well before being sacrificed [254]. In his description of the months of the year, Guaman Poma explains how in the month of March, a black llama is sacrificed [240]. The author clarifies, indicating that even though it is a sacrifice to the idols, nevertheless it is also a time of plenty; there is ample rainfall, which is good for the lands that will be left to rest, and there is an abundance of food for people and animals alike (*Ibid.*I: 171). Similarly, for the month of April, the author notes that red-coloured llamas are sacrificed and it is an occasion for joyous celebration and feasting. Even poor Indians enjoy food and dance, while singing the song of their *puca* or red llamas (*Ibid*). In the month of May, more camelids are sacrificed, while the storehouses are filled to overflowing and llama herders joyously sing "llamaya llamayayn yalla llamaya" (*Ibid*). Taken as a whole, these images and the accompanying text confirm Mercedes López-Baralt's observation that although Guaman Poma criticized idolatrous practices among indigenous peoples, he admired the Incaic imperial apparatus for its efficient organization and exemplary nature (1988: 395). Indeed, llama sacrifices become a sign of prosperity and wellbeing for all.

According to Jonathan Sawday, "in medicine [and, we could add, in veterinary medicine], anatomization takes place so that, in lieu of a formerly complete 'body,' a new 'body' of knowledge and understanding can be created. As the physical body is fragmented, so the body of understanding is held to be shaped and formed..." (Sawday 1995: 2). The three examples analyzed in this chapter have attempted to outline some of the ways in which dissection and anatomization functioned as apt metaphors for expressing the dynamic of Andean-European colonial relations and knowledge production during the eighteenth century. While it can be argued that dissection as well as preservation were key to both European and Andean knowledge systems, it is clear that they carried very different implications. Feuillée's dissection of the guanaco marked a point of transition defined by processes of cutting and transplanting. Specifically, indigenous peoples were excised from the scene of dissection (read: scene of knowledge production) while the body of the animal was claimed by the entwined discourses of science and profit; they were entwined because, as observed earlier, the dissection was initially motivated by the hope of uncovering precious bezoar stones in the animal's gut. In the case of the Fragonard llama, the animal was transplanted to the veterinary school where it was ultimately dissected and preserved as a visual-material archive to the production of knowledge. Mapping the body interior with the red and blue dyes of the injected veins and arteries facilitated the processes of memorization and learning, thereby making the animal body more readily accessible and useful to future scientific and economic projects of acclimatization.[40]

The third example also reflects an implicit colonial context but it stands in marked contrast to the previous two where the indigenous presence was removed. In this final illustration, Guaman Poma's representation of the Quechua sacrifice of a camelid is visually striking for the way it captured the ritualized ceremony at the precise moment when the man's hand reached into the body of the animal struggling at the brink of death. The performative act of reading this rebellious body—rebellious because it does not want to relinquish its hold on life—shifts its signification away from ensuing lifelessness and decomposition and toward the active transmission of memory and the generation of traditional knowledge. For indigenous pastoral peoples, scrutiny and interpretation of the camelid's internal organs were not linked to the desire for profit in the Western sense. Rather, as depicted by Guaman Poma, the ritual emphasized how the culture of dissection was central to vital social memories and practices because it rallied the entire pastoral community, men and women, young and old. In this context, dissection and anatomization reiterated the fundamental bond between humans and camelids, and the knowledge generated by the extraction and reading of the animal's organs underscored the connection between the past and present condition of the animal and the future survival of the community.

Notes

[1] Earlier versions of this essay were presented at the Ohio State University as part of the seminar "Gender and Ethnicity Across the Divides," organized by Ileana Rodríguez; at the conference "Animals and Science: Anthropological Approaches," held in 2005 at the University of Manchester; and at the Nineteenth Annual Debartolo Conference in Tampa, Florida in 2005. I would like to express my appreciation to Reed Benhamou, Nina Berman, Maggie Bolton, Cathrine Degnen, Penelope Dransart, Geraldine Friedman, Nancy Peterson, Ileana Rodríguez, Aparajita Sagar, Barry Shank, and Fernando Unzueta for their comments and suggestions, which have strengthened the essay in countless ways. I am also very grateful to Christophe Degueurce, Professor of Anatomy at the École Vétérinaire Maisons-Alfort and Curator of the Fragonard Museum. Conversations with him regarding the "Fragonard llama" have been illuminating. Special thanks to Tom Stephenson for his help in preparing the photographs of the Fragonard llama.
[2] For example, the Scottish anatomist John Hunter (1728-1793) noted while writing about the dissection of the male sexual organs in his "Observations on Physiology" that generative secretions could be tested by taste: "The semen would appear, both from the smell and taste, to be a mawkish kind of substance; but when held some time in the mouth, it produces a warmth similar to spices, which lasts some time" (Hunter 1861: 189); similarly, "the fine transparent mucus of the urethra is strongly impregnated with sea-salt, which is immediately known by the taste" (*Ibid*: 189). Moore (2005) similarly points to the attention that Hunter gave to anatomy as a multi-sensorial art (79-80).
[3] My reading of knowledge as a form of gain draws from Georges Van den Abbeele (1992), especially where he describes the economy of travel as presenting the possibility of either loss or gain (xvi-xviii).
[4] "Le ministre qui auroit contribué à enrichir le Royaume d'un animal aussi utile, pourroit s'en applaudir comme de la conquête la plus importante." Archives of the Muséum National d'Histoire Naturelle, Paris. MS 864 IIID.
[5] The phrase "'surgical strategies'" is Stafford's (1991: 64). John Hunter's famous experiments transplanting a human tooth into a cock's comb and a rooster's testicle

into a hen's stomach can be read as examples of some of the kinds of scientific inquiries taking place at the time. Other examples include the transplanting of plants and animals from one region, country, or continent to another for the purposes of acclimatization and naturalization.

[6] The term "contact zone" is Pratt's (1992).

[7] This is not to suggest that there were not other accounts where a camelid was opened and examined. Feuillée, however, was the first known Westerner to dissect the body and methodically describe the details of the animal's internal structure.

[8] "Je jugeai par la route qu'ils tenoient, qu'ils passeroient près de ma tente; j'y retournai promptement, desirant les voir de près pour les bien examiner. Je sçavois par le rapport qu'on m'en avoit fait, que leur figure étoit tout à fait extraordinaire, & je souhaitois en faire un dessein" (Feuillée 1725: 21).

[9] On the debates between Feuillée and Frezier see Cañizares-Esguerra (2001: 15-17).

[10] No mention is made of who gave the captain the animals.

[11] On the importance of the bezoar stone for commerce, see the *Dictionnaire Universel de Commerce*, Vol. 1 (1741: 478; 492). See also Stephenson (2010).

[12] On 10 February 1790, one of the zoologists from Alejandro Malaspina's expedition, Antonio Pineda, also dissected a llama while in Panama but his description of the animal focused on its external characteristics and his anatomy on the weights and measurements of the organs. Archives, Museo Nacional de Ciencias Naturales. Alejandro Malaspina's *Expedición alrededor del mundo, 1789-1795.* 598 [1790]. Zoología Panamense. [Antonio Pineda]. María de los Ángeles Calatayud, *Catálogo de las expediciones y viajes científicos españoles a América y Filipinas (siglos XVIII y XIX).* Fondos del Archivo del Museo Nacional de Ciencias Naturales. Madrid: Consejo Superior de Investigaciones Científicas; Museo Nacional Ciencias Naturales, 1984.

[13] See for example Robert Knox, "Observations on the Structure of the Stomach of the Peruvian Lama; to which are prefigured Remarks on the Analogical Reasoning of Anatomists, in the Determination à priori of Unknown Species and Unknown Structures." *Transactions of the Royal Society of Edinburgh.* Vol. XI (1831): 479-498, 3 plates.

[14] Pratt notes that the erasure of non-European humans is also characteristic of Humboldt's work (Pratt 1992: 125).

[15] "Après l'examen de l'oesophage, j'examinai fort soigneusement le premier ventricule, appellé la pance: elle est dans ces animaux d'une grosseur étonnante; j'avois apris des Indiens, que c'étoit dans cet endroit où se formoit ordinairement le Bezoard, mais après une exacte recherche, je n'y trouvai qu'une grande quantité d'alimens très-mal digerés" (Feuillée 1725: 28).

[16] My discussion of the mapping of the body's interior draws from Sawday (1995: 23-25).

[17] Javier Portús called attention to a similar shift in eighteenth-century pictorial representations of the camelids. Prior to this century, images of the animals customarily depicted them in human and ethnographic environments, yet during the eighteenth century they increasingly appeared in isolation from any specific contexts. Portús suggested that this transformation was due to the camelids'

diminishing role in most of Peru and Chile in comparison to the growing importance of imported European animals (Portus 1994: 66).

[18] For more on the history of transporting live animals during the eighteenth century, see Robbins (2002: 9-36).

[19] By the nineteenth century, survival rates for camelids improved somewhat for the transatlantic crossing. My research indicates that even so the animals did not flourish once they reached Europe. See also Penelope Dransart, this volume.

[20] See Simon (2002) for a detailed description of Fragonard's technique of preservation.

[21] For more on the relationship between dissection of the body and colonialism, see Sawday (1995: 25-27).

[22] Most scholars who have studied the anatomist's life and work believe that he was dismissed by the head administrator of the school, Claude Bourgelat, due at least in part to the intense rivalry and jealousy that the latter felt toward Fragonard. See Simon (2002: 68; 2008: 149); Degueurce (1999a); Ellenberger (1981).

[23] Jonathan Simon notes that during the eighteenth century "anatomical preparations ran the whole spectrum of permanence, from temporary accompaniments to particular dissections that illustrated the mechanics of the human body by, for example, injecting fluids into the arterial system, to permanent preparations frozen in time, such as those produced by Fragonard. These permanent preparations made from dead bodies shared at least two advantages with their more expensive wax counterparts. First, they could serve as a permanent display of anatomy to be consulted at any time, and second, they did not smell or decompose and so could be conveniently incorporated into the private natural history cabinets of the anatomist's more refined clients" (2002: 66-67).

Fragonard was not only a skilled anatomist but also a prolific one. In 1783, he returned to Alfort and, with Flandrin, a former student, drew up an inventory of the anatomical collection, which amounted then to 3,033 items (Archives Nationales. F/10/1294. Instruction Publique. Inventaire du Cabinet d'Anatomie de l'École Vétérinaire Nationale d'Alfort. 1783). Unfortunately, only a small number of Fragonard's anatomical creations have survived to the present. For more on the history of the Fragonard museum, see Langlois and Lévesque (1804), Railliet and Moulé (1908), Poulle-Drieux (1962), Chaton (1970), Degueurce (1999a; 1999b) and Simon (2002; 2008).

[24] Indeed, the "Fragonard llama" was most likely prepared by one of Fragonard's students rather than Fragonard himself, because the llama did not arrive at the veterinary school until after the anatomist's dismissal.

[25] Simon argues that Fragonard's work cannot be described as "*both* art *and* science. Rather, it is a reminder that the distinction between art and science in the context of anatomical preparations really became clear only in the course of the nineteenth century" (2002: 76). For more on the intersections of art and science in the study of picturesque anatomy, see Callen (1997).

[26] My reading of the appropriation of the body through the convergence of science and colonialism was informed by Sawday (1995). See especially pp. 25-27.

[27] Sue's most important works include: *Anthropotomie, ou l'Art d'Injecter, de Disséquer, d'Embaumer et de Conserver les Parties du Corps Humain, &c.*

(1765); *Abrégé de l'anatomie du corps de l'homme, avec une méthode sur la manière d'injecter et de préparer les parties fraîches ou sèches* (1748); *Élémens d'anatomie, à l'usage des peintres, des sculpteurs et des amateurs, ornés de quatorze planches en taille-douce, représentant au naturel tous les os de l'adulte et ceux de l'enfant du premier âge, avec leur explication* (1788).

[28] Uvedale Price, in his *An Essay on the Picturesque* found autumnal hues to be especially picturesque, including the "various gradations in the tints of broken ground, and of the decayed parts in old trees; the weather stains, and many of the mosses on stones and trunks of trees; with a thousand more, equally distinct from those which are beautiful. If to these be opposed the soft and tender colours of the stems of young trees, the fresh greens of spring, both in trees and herbage, its flowers and blossoms, it will shew in how many instances picturesque colours as well as forms, arise from age and decay" (1796: 183-184).

[29] Sournia suggests as well that the overinflation of the vessels was prepared deliberately: "Regardons cette présentation d'une face écorchée, où de grosses veines superficielles gonflées d'un liquide bleu donnent à ce visage une allure crépusculaire. Or les autres pièces prouvent que Fragonard savait injecter à des pressions différents les systèmes artériel et veineux; ici cette distension anormale fut voulue, et l'effet cherché fut obtenu" (1981: 5).

[30] As Paul Carter puts it, "picturesque prospects were ones that allowed the eye to wander from object to object: they had their mental counterparts in complex ideas that sparked off trains of thought. Picturesque views might give rise to all kinds of pleasant ideas, but the primary pleasure they gave resided in their picturesqueness itself--in the fact that their structure of casually interlinked and contrasting forms enticed the eye (and the mind) to wander" (Carter 1988: 231-232).

[31] Similar to Carter, Ann Bermingham writes that in Britain, "as several art historians have shown, the aesthetic discovery of landscape masked both the economic transformation of rural nature through the parliamentary acts of enclosure and the political agendas that attended it" (Berminghan 2000: 77).

[32] For a different angle on the relationship between animals, humans, and the concept of possession, see Penelope Dransart in this volume.

[33] See also Dransart 2002.

[34] George Robert Miller's research has shown that in addition to this form of sacrifice, known as *ch'illa*, some Andean communities slaughtered camelids by means of a ventral throat slit. Based on the ethnoarchaeological evidence he collected as well as on his reading of colonial documents, Miller hypothesizes that the ch'illa method of slaughter may have been limited to Quechua-speaking communities since to date there is little evidence of its practice in Aymara territory. See Miller (1979: 25-39). Thanks to Penelope Dransart for bringing Miller's work to my attention.

[35] Penelope Dransart as similarly observed how the nature of the pastures and care of the animals impact directly on the quality of the fiber (2002: esp. 89-94). See also Denise Arnold and Juan de Dios Yapita (2000: 34). Drawing from colonial sources, John Murra observes that llama entrails would also be examined for special omens as part of general war preparations (1965: 207).

[36] Stephanie Bunn in this volume also describes the network of human relations formed in Kyrgyzstan around the detailed preparation and presentation of the slaughtered animal for ritual feasts.
[37] When referring to Guaman Poma's drawings, numbers in square brackets refer to those in the original manuscript. Roman numerals in references refer to volumes of the published edition published in Caracas by Biblioteca Ayacucho.
[38] Guaman Poma adopts Spanish terminology here, referring to the llama as a ram or sheep. Tom Cummins has suggestively argued that by framing indigenous cultural production in Christian images and language – i.e. utilizing the word "sheep" instead of "llama," Guaman Poma is able to communicate certain ideas and images without revealing too much of Andean signifying practices: "Did Guaman Poma understand such practices merely as the actions of artisans, devoid of any deeper meaning? To suggest an answer, it is important to remember that as Guaman Poma composed the *Nueva corónica i buen gobierno*, the description and analysis of Andean culture was not a neutral intellectual act. Any revelation had the possibility of having destructive consequences in terms of extirpation and exploitation. Guaman Poma chronicles some of these destructive acts and mentions by name some of the most important protagonists who acted against Andean cultural forms" (Cummins 1997: 253).
[39] Created sometime during the first and second century A.D., the sculpture is currently displayed in the Louvre with the title "Vieil homme écorchant un animal. Ecorché rustique."
[40] Sawday has described the process in the following terms: "The task of the scientist was to voyage within the body in order to force it to reveal its secrets. Once uncovered, the body-landscape could be harnessed to the service of its owner. In thus establishing the body as 'useful'—a key term amongst the natural philosophers of the seventeenth century—we are able to perceive the language of colonialism and the language of science as meshing into one another...To say of the New World that it was like a body...was to explore a metaphor, but it was also to appropriate the body in the service of colonial expansion" (25).

Bibliography

Abercrombie, T. A. 1998 *Pathways of Memory and Power: Ethnography and History Among an Andean People*. Madison: University of Wisconsin Press.

Adorno, R. 1990. "The Depiction of Self and Other in Colonial Peru". *Art Journal*. 49 (2): 110-118.

—. 1979 "Icon and Idea: A Symbolic Reading of Pictures in a Peruvian Indian Chronicle". *Indian Historian*. 12 (3): 27-50.

Algarotti, F. 1769. *Essai sur la Peinture, et sur l'Academie de France, établie à Rome*. Trans. M. Pingeron. Paris: Chez Merlin.

Arnold, D Y.and J. de D. Yapita 2000 *El rincón de las cabezas: Luchas textuales, educación y tierras en los Andes*. La Paz: Facultad de Humanidades y Ciencias de la Educación, UMSA; ILCA.

—. 1998 *Río de vellón, río de canto: Cantar a los animales, una poética andina de la creación*. La Paz: Facultad de Humanidades y Ciencias de la Educación, Carrera de Literatura, UMSA; HISBOL.

Bermingham, A. 2000 *Learning to Draw: Studies in the Cultural History of a Polite and Useful Art*. New Haven and London: Yale University Press.

Blin, P.C. 1981 "Les écorchés de Fragonard". *L'autre Fragonard*. Paris: Jupilles.

Boitard, P. 1842 *Le Jardin des Plantes. Description et Moeurs des Mammifères de la Ménagerie et du Muséum d'Histoire Naturelle*. Paris: J.-J. Dubochet et Ce, Éditeurs, Rue de Seine, 55.

—. 1835 *Manuel du Naturaliste Préparateur, ou l'Art d'Empailler les Animaux et de Conserver les Végétaux et les Minéraux*. 3rd. ed. Paris: À la Librairie Encyclopèdique de Roret.

Buffon, G. L. L., comte de. 1782 *Supplément à l'Histoire Naturelle, Générale et Particulière*. vol. 6. Paris: Imprimerie Royale.

Callen, A. 1997 "The Body and Difference: Anatomy Training at the Ecole des Beaux-Arts in Paris in the Later Nineteenth Century". *Art History*. 20 (1): 23-60.

Cañizares-Esguerra, J. 2001 *How to Write the History of the New World: Histories, Epistemologies, and Identities in the Eighteenth-Century Atlantic World*. Stanford: Stanford University Press.

Capus, G. 1883 *Guide du Naturaliste Préparateur et du Voyageur Scientifique, ou Instructions pour la Recherche, la Préparation, le Transport et la Conservation des Animaux, Végétaux, Minéraux, Fossiles, et Organismes Vivants, et pour les Études Histologiques et Anthropologiques*. Deuxième Édition entièrement refondue par le Dr. A.T. de Rochebrune, Aide-Naturaliste au Muséum d'Histoire Naturelle. Paris: Librairie J.B. Baillière et Fils.

Carter, P. 1988 *The Road to Botany Bay: An Exploration of Landscape and History*. New York: Alfred A. Knopf.

Chabert, P., J.-B. Huzard, and Flandrin. 1782-1790 *Instructions et Observations sur les Maladies des Animaux Domestiques; Avec les moyens de les guérir, de les préserver, de les conserver en santé, de les multiplier, de les élever avec avantage, & de n'être point trompé dans leur achat. On y a joint l'analyse des ouvrages vétérinaires, anciens & modernes, pour tenir lieu de tout ce qui est écrit sur cette science. Ouvrage nécessaire aux Cultivateurs, aux Proprietaires de Bestiaux, &*

aux Artistes Vétérinaires. 3rd ed. À Paris, De l'Imprimerie & dans la Librairie Vétérinaire de la Citoyenne HUZARD, rue de l'Éperon, No. 11, quartier St.-André-des-Arts. An VII.

Chaton, M. 1970 *L'Epoque Alforienne de Claude Bourgelat et la Création de l'Ecole Vétérinaire d'Alfort.* Thèse pour le Doctorat Veterinaire. no. 109. Paris: Interbus.

Cummins, T. 1997 "Images on Objects: The Object of Imagery in Colonial Native Peru As Seen Through Guaman Poma's *Nueva Corónica i Buen Gobierno*". *Journal of the Steward Anthropological Society.* 25 (1-2): 237-273.

Cuvier, G. 1835 *Leçons d'Anatomie Comparée.* 2nd ed. Vol. 4. Part 2. Paris: Crochard et cie..

Degueurce, C. 1999a. "Fragonard (1732-1766). Un Obscur au Siècle des Lumières". *Presentation 27.* http://musee.vet-alfort.fr/Site_Fr/SFHMSV/SFHMSV_files/Textes/Activites/Bulletin/Txts_Bull/B1/Degueurce_B1.pdf

Degueurce, C. 1999b "Le Musée Fragonard, École Nationale Vétérinaire d'Alfort". In G. Tilles and D. Wallach (eds.) *Les Musées de Médecine: Histoire, Patrimoine et Grandes Figures de la Médecine en France..* Toulouse: Éditions Privat: 91-104.

Dictionnaire Universel de Commerce: Contenant Tout ce qui Concerne le Commerce qui se Fait Dans les Quatre Parties du Monde.... Par M. Philemon Louis Savary, Chanoine de l'Eglise Royale de S. Maur des Fossez, son Frere. Nouvelle Edition. Tome III. À Paris, Chez la Veuve Estienne, rue Saint Jacques, à la Vertu, 1741.

Dransart, P. 2002 *Earth, Water, Fleece and Fabric: An Ethnography and Archaeology of Andean Camelid Herding.* London, Routledge.

Dyce, K.M., W.O. Sack, and C.J.G. Wensing. 2002 *Textbook of Veterinary Anatomy.* 3rd ed. Philadelphia, London, New York, St. Louis, Sydney, and Toronto: Saunders.

Ellenberger, M. 1981 "L'autre Fragonard. Essai Biographique". *L'autre Fragonard.* Paris: Jupilles.

Feuillée, L. 1725 *Journal des Observations Physiques, Mathematiques et Botaniques, Faites par Ordre du Roi sur les Côtes Orientales de l'Amerique Méridionale, & aux Indes Occidentales. Et dans un autre Voïage fait par le même Ordre à la Nouvelle Espagne, & aux Isles de l'Amerique.* Tome III. Paris: Chez Jean Mariette.

Flores Ochoa, J. A. 1977 "Enqa, enqaychu, illa y khuya rumi". In *Pastores de puna: Uywamichiq punarunakuna.* Ed. Jorge A. Flores Ochoa. Lima: Instituto de Estudios Peruanos.

Garcilaso de la Vega, Inca. 1985 *Comentarios reales de los Incas*. Ed. Aurelio Miro Quesada. Vols. 1 and 2. 2nd. ed. Caracas: Fundación Biblioteca Ayacucho.

Gilpin, W. 1792 *Three Essays: On Picturesque Beauty; on Picturesque Travel; and on Sketching Landscape: to which is added a Poem, on Landscape Painting*. London: Printed for R. Blamire, in the Strand.

Guaman Poma de Ayala, F. 1980 *Nueva corónica y buen gobierno*. Ed. Franklin Pease. Vols. 1 and 2. Caracas: Biblioteca Ayacucho.

Haraway, D. 1989 *Primate Visions. Gender, Race, and Nature in the World of Modern Science*. New York and London: Routledge.

Hunter, J. 1809 "Directions, for Preserving Animals, and Parts of Animals for Anatomical Investigation; and Concerning Extraneous Fossils." London: J. Adlard, National Library of Medicine. History of Medicine Collection. W6 P3. v. 7481. box 354. no. 13.

Hunter, J. 1861 *Essays and Observations on Natural History, Anatomy, Physiology, Psychology, and Geology*. Ed. Richard Owen. Vol. 1. London: John Van Voorst.

Jordanova, L. J. 1985 "Gender, Generation and Science: William Hunter's Obstetrical Atlas". In W.F. Bynum and Roy Porter (eds.), *William Hunter and the Eighteenth-Century Medical World*. Cambridge, London, and New York: Cambridge University Press: 385-412.

Langlois, N.G., and J.F.O. Lévesque. 1804 *Notice Descriptive de l'École Vétérinaire d'Alfort, Contenant les Conditions de l'Admission des Élèves, leur Entretien, un Apperçu de l'Enseignement, Et la Description du Cabinet*. Paris, de l'Imprimerie de Marchant, Collège d'Harcourt, Rue de la Harpe, Vendémiaire An XIII.

López-Baralt, M. 1988 *Icono y conquista: Guamán Poma de Ayala*. Madrid: Hiperión.

MacLeod, R. 1988 "Introduction". In R. MacLeod and M. Lewis (eds.), *Disease, Medicine, and Empire: Perspectives on Western Medicine and the Experience of European Expansion*. London and New York: Routledge.

Miller, G. 1979 *An Introduction to the Ethnoarchaeology of the Andean Camelids*. PhD Dissertation, University of California, Berkeley. Ann Arbor: University Microfilms.

Moore, W. 2005 *The Knife Man: The Extraordinary Life and Times of John Hunter, Father of Modern Surgery*. London, Toronto, Sydney, Auckland, and Johannesburg: Bantam Press.

Murra, J. V. 1965 "Herds and Herders in the Inca State". In A. Leeds and A. Vayda (eds.), *Man, Culture, and Animals: The Role of Animals in*

Human Ecological Adjustments. Publication no. 78. Washington, D.C.: American Association for the Advancement of Science.

Portús, J. 1994 "Europe and the Andean Camelids: Words and Pictures Tell a Story of Cultural Integration." In J. Flores Ochoa, K. MacQuarrie and J. Portús (eds.), *Gold of the Andes: The Llamas, Alpacas, Vicuñas, and Guanacos of South America*. Trans. Joanna Martinez. Vol. 2. Barcelona: Francis O. Pathey and Sons.

Poulle-Drieux, Y. 1962 "Honoré Fragonard et le Cabinet d'Anatomie de l'École d'Alfort pendant la Revolution". *Revue d'Histoire des Sciences et de Leurs Applications*. 15: 141-162.

Pratt, M. L. 1992 *Imperial Eyes: Travel Writing and Transculturation*. London and New York: Routledge.

Price, U. 1796 *An Essay on the Picturesque, as Compared with the Sublime and the Beautiful; and, on the Use of Studying Pictures, for the Purpose of Improving Real Landscape*. Vol. 1. London: J. Robson.

Railliet, A., and L. Moulé 1908 *Histoire de l'École d'Alfort*. Paris: Asselin and Houzeau.

Robbins, L. E. 2002 *Elephant Slaves and Pampered Parrots: Exotic Animals in Eighteenth-Century Paris*. Baltimore: The Johns Hopkins University Press.

Rothfels, N. 2002 *Savages and Beasts: The Birth of the Modern Zoo*. Baltimore and London: The Johns Hopkins University Press.

Sawday, J. 1995 *The Body Emblazoned: Dissection and the Human Body in Renaissance Culture*. London and New York: Routledge.

Simon, J. 2008 "Honoré Fragonard, anatomical virtuoso". In B. Bensaude-Vincent and C. Blondel (eds.), *Science and Spectacle in the European Enlightenment*. Aldershot: Ashgate.

—. 2002 "The Theater of Anatomy: The Anatomical Preparations of Honoré Fragonard". *Eighteenth-Century Studies*. 36.1: 63-79.

Sournia, J.-C. 1981 "L'Anatomie du Tragique". *L'autre Fragonard*. Paris: Jupilles: 4-5.

Stafford, B. M. 1991 *Body Criticism: Imaging the Unseen in Enlightenment Art and Medicine*. Cambridge, MA and London, England: The MIT Press.

—. 1984. *Voyage Into Substance: Art, Science, Nature, and the Illustrated Travel Account, 1760-1840*. Cambridge, Massachusetts, and London: The MIT Press.

Stephenson, M. 2010 "From Marvelous Antidote to the Poison of Idolatry: The Transatlantic Role of Andean Bezoar Stones during the late Sixteenth and Early Seventeenth Centuries." *Hispanic American Historical Review* vol. 90 no. 1: 3-39.

Turgot, É.-F. 1758 *Mémoire Instructif sur la Manière de Rassembler, de Preparer, de Conserver, et d'Envoyer les Diverses Curiosités d'Histoire Naturelle; Auquel on a joint un Mémoire Intitulé: Avis pour le Transport par Mer, des Arbres, des Plantes Vivaces, des Semences, & de Diverses Autres Curiosités d'Histoire Naturelle.* Paris: Chez Jean Marie Bruyset.

Van den Abbeele, G. 1992 *Travel as Metaphor: From Montaigne to Rousseau.* Minneapolis and Oxford: University of Minnesota Press.

CHAPTER TWO

ANIMAL PLAGUE: CULTURE, ORDER AND DISEASE IN LATE VICTORIAN BRITISH DAIRY FARMING

RICHIE NIMMO

> "Disease is at one and the same time disorder–the existence of a perilous otherness within the human body, at the very heart of life–and a natural phenomenon with its own constants, resemblances, and types."
> (Foucault 1970: xxiv)

Introduction

In this chapter I undertake a genealogy of humanist discourse and practice in the British dairy industry of the late nineteenth century. According to this discourse human beings inhabit an ontologically unique domain incommensurable with the nonhuman world; this domain is what modernity knows as "culture", an exclusively human world of subject-agents and inter-subjective meanings. My central argument is that this category of "culture" at the organising heart of modern ontology, far from being the given condition of human beings, has been historically constituted through the discourse of humanism and the material and epistemological practices in which its categories are actualised. At the core of this is the attempt to show that what we think of as "culture", hence how we experience "culture", that is, "culture" as a mode of organisation of experience, is produced through networks of historical practices inscribing what it is to be human. This presupposes that the historically specific ways in which we categorise experience and classify knowledge lead to historically specific divisions of experience itself, that experience is not an unmediated, a-historical phenomenon.[1] The corollary is that there is no culture either separate from or preceding "culture"; culture is not the metaphysical moment of presence that ethnographic methods sometimes imply.

Until fairly recently in the social sciences the terms “culture” and “human” were generally used in a manner that was deeply tautological; to be human was to be a cultural being, just as it was taken for granted that to be part of the world of culture was to be human.[2] Where the term “social” was preferred the same assumption was present: humans were social beings, and to be social was to be human. The conceptual slippage went unrecognised for the two terms were regarded as synonymous. Hence modern social science is very much the creature of humanist ontology; its conception of its own object of knowledge, its notion of the “social/cultural”, has been profoundly shaped by and still largely remains in thrall to the ontology of humanism. I do not suggest that this discourse is somehow exclusively modern, for its antecedents clearly stretch back into antiquity; but what I am calling “late” modernity sees its realisation in a multitude of practices and institutions and its operationalisation in scientific and technical knowledges, whereas previously it had been largely confined to the abstract discourses of theology, philosophy, and latterly literature.[3] In the second half of the nineteenth century humanism began to become materially invested within a whole complex of everyday practices and technologies; rather than a reflective abstraction, the “man” of modern humanism became the concrete production of an elaborate politico-epistemological apparatus.

In substantiating this argument I focus upon the work of humanism within a relatively circumscribed set of historical practices, which are illustrative of a more general, though not necessarily wholly concurrent or identical, process. My thesis is that the period from 1865 to around 1900 in dairy agriculture saw an unprecedented entrenchment and actualisation of humanist discourse through the network of political technologies which were deployed under the rubric of sanitary discourse-practice and the nascent sciences of microbiology and veterinary pathology. This realisation of humanist discourse was intimately connected to the struggle against diseases of cattle, to the expansion of the milk trade and the associated drive for clean milk, and crucially to the battle against diseases transmissible to humans through dairy milk. I contend, moreover, that this process was closely bound up with the rationalisation of practices and institutions integral to “late” modernity or modernity proper. Though I suggest that their common matrix was a modernity constituted in and through humanist discourse, I do not however try to address the question of the relations between the field of dairy agriculture and other fields of practice in any systematic way here, not least because it is an intricate historical problem and beyond the scope of this paper. The argument I make is grounded in an analysis of historical documents and secondary source materials. My primary sources consist of two national agricultural journals, the Journal of the Royal Agricultural Society of England (RASE) and the Journal of the British Dairy Farmers' Association (BDFA), which I have studied in detail for the period in question. As well as using these empirically for the historical information which they contain concerning dairy and veterinary practices, I have also sought to trace and to problematise the organising categories and structures of knowledge which they embody, and to situate these genealogically in order to grasp their role in the historical constitution of subjects, objects and domains.[4]

The paper is organised into three parts: The first outlines the main theoretical dimension of the argument; it considers the ontological significance of disease and the nature of its threat to the humanist categories of modernity by introducing the concept of the zoonose. This section also clarifies my understanding of humanist discourse, its instruments and disciplinary technologies, and sets out my principal differences with Bruno Latour and actor-network theory, from which I draw inspiration and with which I share some common themes and analyses. The second part commences the properly historical and empirical substance of the paper. It comprises an analysis of the conjuncture in which agricultural statistics and experimental microbiology gave rise to the aggregation and politicisation of animal disease in the wake of the 1865 rinderpest or cattle plague, and which laid the ground for the interventions of sanitary discourse and veterinary pathology. The third section contrasts the fluid and hybrid spatiality of the zoonose with the analytical space produced by sanitary science and practice, which sought to re-impose the dualist categories of humanism upon a disordered zoonotic space. It considers the progress of sanitary legislation against robust opposition, and crucially how the tension between sanitary imperatives and economic exigencies was rooted in their contradictory principles for the use of space. Further exploring the relationship between humanist and zoonotic spatiality, this section then looks at the debate on cattle housing and the design of cattle sheds, showing how the animal body and its immediate environment was broken up by sanitary discourse into discrete functional parts, and how this was facilitated by an ontological distinction between what was inside the animal and what was outside. I conclude by bringing together the key developments and by suggesting how the opening arguments may be borne out by the intricacies of actual history.

Zoonoses: An Ontological Disorder

Disease is usually thought of as a thoroughly natural phenomenon, properly the object of biological science and pathology. Insofar as it is allowed that disease has a social dimension, this is strictly separated from its biological, natural aspects. On the one hand we have the sociology of health, on the other hand epidemiology, each hermetically sealed into its own conceptual domain. Thus one may study at will the relationship between health and social class, analyse the changing discourses around disease and its prevention, or study the effects of particular policies, but as soon as one entertains the notion that disease *itself* may be a social phenomenon, rather than merely a natural one which is relationally distributed, discursively understood and politically managed in deeply social ways, then one quickly encounters the conceptual limit of social science, the self-imposed horizon of its dualist imagination.[5] But what could be more social than disease? Disease is every bit as collective as language, indeed more so, for it speaks the universal language of matter, recognising no linguistic boundaries and excluding no body from its community.

If disease constitutes a social relationship, quite over and above the social relations *of* disease, then it is a relationship that profoundly challenges our inherited conception of the social. It is a material relation between bodies, but one which elegantly undermines the Cartesian notions of human meaning and agency that still so often underpin sociological discourse *on* the body, leaving us with an inter-corporeality that is still stubbornly social but which refuses to be reckoned back into some form of idealist, linguistic, semiotic, in short narrowly "cultural" conception *of* the social. This is distinct from the notion of disease as *metaphor*, which is one way that the sociality of disease has been conceptualised by medical anthropologists.[6] My point is not that disease has cultural and metaphorical significations, a claim which hardly upsets the customary dualism of social science, but that disease *itself*, in its very biological dynamics, is thoroughly and immediately social. Thus understood, disease compels us to conceptualise sociality without tacitly circumventing recognition of its inherent materiality in materialist terms. At a stroke sociality seems to rise up before us and cast off the dualist categories which for so long we had relied upon to contain it. Disease then is deeply problematic, for it fundamentally transgresses the great divide between mind and body, subject and object, culture and nature, human and nonhuman, which is the ontological heart of modernity, its knowledges and disciplines. More than just a practical threat to human health therefore, disease poses a threat, at once material and ontological, to modernity, its idea of "culture", indeed its very notion of "humanity".

No kind of disease more dramatically destabilises the dualist categories of modern humanism than a disease that is communicable between animals and humans–a zoonose. In its transmission from nonhumans to humans the zoonose threatens the ontological security of the human domain by its blurring of the anthropocentric boundary between "culture" and "nature".[7] The very materiality of the zoonose undermines the separation of an exclusively human domain from the nonhuman world. It is this separation that effectively constitutes "culture" in modern ontology, and constitutes it as the secular translation of a divine or sacred realm, in which human action and meaning hold sway and from which history is born. Indeed just as the modern secular sphere emerged from thoroughly religious roots, so this notion of "culture" involves what is ultimately a theological conception of humanity.[8] By divorcing "culture" from "nature" and claiming it exclusively for humans we have managed to tacitly retain the notion that human lives are essentially affairs of the spirit, that humans, in other words, have "souls". It then becomes possible to contrast human lives with the purely bodily, material lives of nonhuman animals, which must live outside of the sanctified sphere of "culture" (Foucault 1991b: 29-30). Just as nonhumans were denied entry to the community of souls in the neo-Aristotelian Catholic theology of St. Thomas Aquinas, so they are denied entry to "culture"–that modern community of souls–in the secular theology of humanism.[9]

Animals nevertheless continue to occupy a marginal and therefore dangerous position in relation to humanism for they are inherently hybrid, refusing to be either of "nature" or "culture", and thus they present a unique challenge to the coherence and purity of these meticulously constructed categories.[10] It is not merely that animals fail to fit neatly into these binary domains and so constitute an anomalous set of cases, leaving the categories intact, rather they show us the contingent nature of the categories themselves. In animals we recognise an other-worldliness which is nevertheless a kind of "subject" like ourselves; whilst so far from being "human", animals are at the same time so close to being "human" that they seem to pose an existential question not merely as to the boundaries of "humanity", but also to our deeply held notions of its value (Berger 1980: 4-5). Nonhuman animals are therefore an irresolvable problematic which humanist discourse must suppress and contain. Only by rendering living animals part of an object-nature, hence purifying the human domain of subject-culture, in both discourse and practice, can the binary ontology of humanism be sustained.

In relation to this ontology the zoonose is the great leveller, proclaiming the radical equivalence of animal being, showing that everything takes place on the same level, and rendering visible the Christian "Great Chain of Being" underpinning secular humanism.[11] Thus the zoonose is a dangerous hybrid, an ontological disorder; it demands a work of purification which is at the same time a work of discipline and an imposition of order. I differ slightly with Bruno Latour, whose notion of the "Modern Constitution" does not consider human-animal relations *as* relations, but sees them rather as the by-product of a modern ontology centred upon a subject/object, society/nature dualism. Thus, like actor-network theory (ANT) more broadly, for Latour "things"–and especially technological objects–tend to take precedence over "beasts" in defining the domain of nonhumanity (Michael 1996: 131). My position turns this on its head by positing human-animal relations as the absent and unstable centre of these ontological dualisms, suggesting that our lived relations with nonhuman animals, as the most perpetually unstable and marginal case for culture/nature dualism, play a critical generative role in the constitution of modern "culture".

Far from being subsumed under a theory of objects, things or technologies then, animals need to be grasped in their specificity and viewed as pivotal in defining the boundaries of the human for modernity. This is why I use the term "humanism" rather than Latour's "Modern Constitution", for I wish to emphasise the *species* relations underpinning the human/nonhuman dualism of modern ontology. Our encounters and relations with nonhuman animal "others", unless meticulously policed by networks of humanist discourse-practices, have the potential to induce destabilising and transformative reflections upon our own "nature" as humans in a way that is very different from our encounters with technology. It is not "cyborgs" which are at the frontier of our encounter with a possible post-human other, but "beasts", "mutants" and "monsters". Thus ANT's typical focus upon technological nonhumans can be displaced by emphasising the critical importance of living nonhumans, which may well have been modified by human agency but are not in any primary sense human "constructs".

Similarly, Latour sees hybrids as proliferating under modernity, hence as somehow *produced*; but if nonhuman *animals* are hybrids, and obviously *not* produced as such, then this has to give way to a view of hybridity as the given state of being, as a condition of existence rather than something historically specific, bound up with objects, and characteristic of modernity. Moreover, the notion that hybrids are produced implies that the raw material for their production is somehow unproblematically non-hybrid, which is in the end an essentialist proposition. It is not that hybrids are produced from a world of essences, but on the contrary, it is essences that must be produced from a world which is always already hybrid. It follows that, rather than Latour's focus upon the relationship between what he conceives as the twin processes underpinning modernity–the proliferation of hybrids and the work of purification respectively – what demands to be explained is the material and discursive production of non-hybrids assigned to either "culture" or "nature", which I understand to be the work of humanist discourse and practice.

Humanism then labours to re-establish the binary divisions which sustain the anthropocentric purity of "culture", in this case by containing the zoonotic disease itself, which threatens to blur such divisions materially. This is not of course the explicit aim of the institutions, knowledges and techniques involved (their conscious purposes are more immediately practical) but the discourse frames such practices as disease control whilst at the same time these practices instantiate its categories. Thus humanism constitutes the overarching regime of truth, the taken for granted organising paradigm, which renders the zoonose not merely an economic problem, nor simply a problem of public health, but simultaneously a theo-ontological logical problem for modernity.[12] Political in the sense that its instruments are those of official regulation, of disciplinary enforcement, of scientific and administrative knowing, humanism operates through forms of knowledge, rationalities, and the management of bodies in space, by constituting objects and domains, defining what can be known, what can act upon it, and by what means. It works "horizontally" across social power relations, constituting their epistemological ground and extending its reign inside their networks of domination.

In another sense the politicality of humanism is unlike what is usually thought of as the political, since its objective is not the domination of one human social group by another, but the definition of what lies within the domain of the human, the social and the cultural, and what does not; it is an "ontological politics" which is the condition of possibility of what is usually thought of as politics. That is why humanism is best grasped as a discourse-practice, for its object is the entrenchment and perpetual reproduction of a specific ontological division of the world, which both constitutes and legitimises a particular mode of species relations. Thus the sanitary discourse and veterinary pathology incipient in the latter part of the nineteenth century, which attempted by an array of increasingly exacting measures to contain animal disease, both zoonotic and epizootic, were the disciplinary elements in an ensemble of technico-epistemological practices serving to police the material and discursive boundaries of the human; taken together these were constitutive of a political technology of humanism. This does not imply that the discourse of humanism is in some way hidden beneath the surface of social life. On the contrary, humanism as I understand it is thoroughly present in social life, open to view and manifest in material forms, made actual through a multitude of practices and knowledges, and so entrenched that its apparent invisibility is the product of its sheer ordinariness–it is the very grammar of the "natural".

The Government of Nonhumans

The first concerted attempt to compile official statistics for agriculture in Britain was not made until 1866. The concept of an aggregate economic loss *to the nation* resulting from cattle diseases only became thinkable once these statistics had begun to be established. Prior to 1866 disease was widely considered an inevitable nuisance to individual farmers, which they had little choice but to accept to a large degree.[13] Indeed, pleuro-pneumonia and foot-and-mouth disease were almost endemic from about 1840, yet no systematic attempt was made to control the spread of contagion. The 1865-66 outbreak of rinderpest or "cattle plague" however and the organisational response it precipitated, led to the emergence of a significant new attitude towards cattle diseases. Agricultural statistics made possible for the first time the concept of a definite animal "population" of a given national territory, to which could be ascribed an aggregate economic value. This meant that for the first time the scale of the financial losses resulting from disease became quantifiable, and the losses of 1866 were quickly condemned as unacceptable (Duckham 1893: 277). The health of nonhumans suddenly became visibly linked to the economic wealth of the nation, in a connection which would have far-reaching consequences.[14] No longer were diseases of cattle everywhere spoken of as an unfortunate fact of agricultural life; instead they were increasingly decried as an entirely preventable national economic problem capable of being solved by rational organisation based on principles drawn from the nascent science of veterinary pathology.

Before 1866 the prevailing view, which had arisen from centuries of practical accommodation to circumstances, was one of fatalistic balance–it held that nature could be manipulated towards human ends only within definite limits and even then only at the cost of contingent re-incursions by nature into what little humans had managed to extract from it; there could be only pyrrhic victories over nature. This now began to be challenged by a distinctly modernist ethos in which nature could be manipulated with precision and its balancing mechanisms thwarted comprehensively by scientific knowledge. Rather than a result of blind forces of nature, or acts of God, the spread of animal disease came to be seen as a consequence of human practice:

> "Until the ravages of plague in 1865-6 led to apprehensions for the safety of the cattle population of this kingdom, the idea of the general adoption of measures for the extinction of disease was never entertained." (Brown 1893: 283)

> "In the period comprised between 1865 and 1893 a complete change has been slowly developed in the agricultural mind in reference to the method of dealing with animal plagues." (Brown 1893: 286)

This shift was underpinned by an epistemic transformation bound up with the early experimental study of bacteria. No longer could disease be viewed as the product of general atmospheric conditions or of heredity, in what had been the dominant theory of "spontaneous generation"; instead it was the specific product of a definite organism, which thrived under certain conditions. Herein was an acknowledgement that nature included unseen "agents" which acted upon the animal body and were the "true cause" of disease, rather than nature being merely an "environment" for agents that could only ever be external to it, whether in the shape of godly or of human intervention. At the same time as nature was admitted a certain "agency", the animal body, its various products and excretions, became a kind of "environment" for the invisible and potentially zoonotic agents in nature.[15] The result was to effectively destabilise the established boundaries between subject and object and to radically decentre the agency which was the defining moment of "culture":

> "Severe as the trial of the cattle plague was, so long as it lasted, it did more to educate the public mind than all the losses that the country had previously suffered amongst its flocks and herds, and induced the study of certain animal diseases that the faculty had previously paid little attention to. The diagnosis of diseases is now better understood, as is also their pathology; the theory of spontaneous generation has been upset and the true origin of many diseases discovered." (Duckham 1893: 281)

> "Prior to these experimental investigations and the discovery of the bacillus by Koch, tuberculosis was supposed to be due to various causes, among which may be mentioned climatic conditions, defective nutrition,

> exposure to hardships, over-work, excessive secretion–in fact, any and every influence admitted to produce debility."
> (Duguid 1890: 309)

The emergent science of pathology centred upon the question of how to effectively intervene in these newly revealed networks of force or "actor networks" in order to contain and manage their problematic "agency". The development of the experimental study of bacteria meant that subject and object were in danger of dissolving into a myriad of diverse connections between unclassifiable things, shorn of their fixed ontological identities; they had to be re-solidified and repositioned in accordance with the binary categories of humanism. This was the work of the very discourse and practice in terms of which the new knowledge was organised. Scientific pathology was therefore the *technical* struggle of humanist discourse against the *ontological* threat of bacteria.[16] It had to increase the material effectiveness of human action to the point where it could negotiate the blurring of subject/object boundaries and definitions by making itself appear qualitatively distinct from the power of action available to invisible nature. Only in this way could the natural "agency" revealed by the microscope be made to disappear once again, by being distinguished from a human agency which appeared to transcend it. If only a quantitative difference of degree existed, this had to be stretched technically until it resembled the qualitative difference of kind demanded by "culture"; human action had to be magnified. It is through such processes that the very notion of human "agency" as a preponderant force acting upon a "nature" outside of itself was made sustainable, by being distinguished from "natural" forces–which were thereby denied the status of "agency"– and set on a plane of "culture" brought into being by this very distinction. Thus the struggle to control cattle disease was far more than an economic issue, for at the same time it was an attempt to maintain a given ontology, an ontological politics, which crystallised a specific form of socio-natural order.

In a conjuncture born of contingent developments in statistics and microbiology then, animal disease became de-individuated; it was no longer the chance misfortune of individual farmers, but instead came to be regarded as a national crisis demanding collective action.[17] This shift could not be realised overnight however; the condition of possibility of the very idea of collective action on animal disease was the extension of a machinery of practices rooted in the newly developing forms of knowledge. Despite the lessons of the 1865-66 cattle plague and the early moves towards centralisation and coordination that it precipitated, the government still lacked an administrative apparatus with sufficient resources even to enable the systematic recording of the incidence of cattle diseases (Duckham 1893: 262). Though only temporarily, for example, the obtaining of official returns for the incidence of foot-and-mouth was abandoned in despair in 1872, revealing the degree to which the foot-and-mouth epizootic was at that time still beyond administrative control, and the extent to which the struggle against cattle diseases was a driving-force for organisational modernisation (*Ibid*: 272). A network of statistical surveillance, regulation and enforcement had to be constructed, not to mention the training and effective coordination of the army of experts who would be needed to apply the new knowledge. This was taken up by the political institutions and became a legislative question; it was at the centre of a raft of parliamentary Bills from 1866 onwards, setting out increasingly stringent, progressively standardised measures to be implemented by structures of local authority, as the various layers of local governance became the means to an end that was increasingly national. Indeed, it was the cattle plague above all which gave rise to the Chambers of Agriculture, marking a significant early step towards centralisation.

This process of incremental rationalisation and standardisation characterised the uneven organisational trajectory of the dairy industry over the next half century, as what came to be seen as the national struggle against animal disease was linked to an emerging project of political and economic centralisation:

> "It became manifest that pleuro-pneumonia would not be exterminated unless it was taken in hand by a central authority and compensation was provided from the national exchequer." (Duckham 1893: 277)

Knowledge of the animal body and the control of animal disease was inextricably a matter of social and political order. Indeed the more traditionally political and industrial questions concerning the social organisation and regulation of production, and scientific questions concerning the pathology and aetiology of cattle disease, were utterly inseparable in practice:

> "Far from being a medical or scientific question, the subject became of as much interest and importance to the stock owner and others connected with the cattle trade, as to the sanitary medical officers and the guardians of the public health." (Duguid 1890: 305)

> "One other important result of legislative supervision has been to drive out of the business the small unscrupulous retailers who lived by fraud, and to throw the great bulk of the trade into the hands of large and responsible companies, such as we see in the Dairy Supply, the Aylesbury, the

> Express, and others. These cannot be regarded otherwise than as institutions of public health, since they are conducted on lines in strict accord with the principles of sanitary science." (Wortley Axe 1892: 62)

Already profoundly theological, animal disease was now rendered political, and it was upon this newly constituted political terrain that sanitary discourse and veterinary pathology framed their interventions. Their object of knowledge was the animal body itself, as pathology sought to determine the most effective means to control disease by understanding in detail its precise corporeal constitution. Thus the bodies of nonhumans came to be identified in a discourse and practice that was at once scientific and political as a dangerous source of pollution and disorder demanding urgently to be controlled.[18] The government of nonhumans became at once conceivable and necessary:

> "In the past, legislation for the prevention of animal plagues appears to have been adopted exclusively in cases of emergency. Only in modern times do we meet with Acts of Parliament dealing with the whole subject, and conferring powers on administrative bodies to be exercised as occasion may demand." (Brown 1893: 281)

Policing the Great Divide

Only the most general legislation intended to prevent the spread of animal diseases existed before the 1865-66 cattle-plague, and it was only sporadically enforced. It focussed upon regulation of the movement of animals and animal products, indicating dawning recognition of the centrality of movement through space, the circulation of bodies and things, commodities and microbes, to the mode of being of the zoonose itself (Brown 1893: 282). The zoonose thrives in dynamic conditions of heterogeneity; its space is the space of flows, the space of mixtures and hybrids, and it was recognised that the movement of cattle to, around, and from markets and fairs was responsible for propagating the disease more widely.[19] This found its nemesis in the fixing, containing, classifying, bounded space increasingly created by sanitary discourse and veterinary pathology throughout the late nineteenth century, which sought to impose ontological order and security through its categorisation, subdivision and analytical dissection of movement. There was a tension between the value of animals as live "stock", as capital, which depended upon their movement to and from locations of trade and exchange and which involved a fluid use of space, and the necessity of imposing rigid, demarcated space upon diseased cattle in order to inhibit the spread of the disease. It is telling in this respect that advocates of free trade denounced the measures necessary to eliminate cattle diseases as a form of protectionism; they perceived that combating foot-and-mouth disease for example, required regulation, inspection, slaughter, and above all the restriction of the movement of "livestock", not least to and from markets.

Far from pathology then being indistinguishable as a political technology of humanism from what might be expected to have resulted in any case from practicalities of capitalist agriculture, in fact there was a significant degree of contradiction between sanitary discourse and organisation on the one hand and immediate economic interests and pressures on the other. This was manifest in the protracted struggle within the industry between the advocates of targeted slaughter as a key means of disease control, and those for whom disease was more or less inevitable and the financial losses incurred by slaughter unacceptable.[20] There was no smooth and universal transition but fierce contestation in which sanitary discourse and scientific pathology eventually won out over other interests.

In the attempt to manage, organise and thereby eliminate the disorderly and transgressive space of the zoonose, not just the animal body but its immediate "environment" was broken-up into its constituent parts, into the various matter and forces which might either inhibit or contribute to the transmission of the disease.[21] Against this background there was a widespread debate as to whether it was better to house cattle in sheds for all or part of the year, or to graze them permanently outdoors, and the question of how cattle sheds should be designed in order to optimise sanitation and inhibit the spread of disease (McFadyean 1895: 658). These specific issues were manifestations of a more general discourse on the management of animal bodies and their material conditions of existence:

> "The object of the designer certainly should be as nearly as possible to obtain walls and floorings which are impervious to the liquid excreted from the animals which, during a great part of the day in some seasons and still a greater part of the night, are obliged to stay in that somewhat confined space. It is also very desirable to have sufficient ventilation. There is a fiction current that 700 or 800 cubic feet of space is wanted for every cow. If that were a necessity, there is probably not one cowshed in a hundred which could be kept open in a large town. But in my opinion the question of cubic space is unimportant in comparison with the provision of means for the plentiful introduction of fresh air. As for drainage... everything should be so arranged that all excreta will either gravitate to the outside, or be easily swept away." (Brown 1887: 48-49)

Not only abstract homogeneous space but also what it contained–the precise quantity of air required for ventilation, calculated in terms of the minimum cubic volume required per cow; the amount of light, assessed in terms of the number and positioning of windows; and the ease and direction of movement of different fluids, both urine, a source of pollution, which was to be encouraged to drain away, and milk, which was to be kept free of all pollution–these became the objects of an accumulative, instrumental knowledge.[22] The formulas adopted were somewhat arbitrary and inconsistent at first, varying considerably between local authorities; they were derived *ad hoc* from the design of army barracks and boarding schools, where similar questions of corporeal government and disciplinary organisation had recently begun to be systematically addressed in relation to human bodies (Foucault 1991b: 142-143):

> "Few reliable tests have been made on animals, but on man we have sufficient data to guide us in the present enquiry. Army barracks are designed on a basis of an allowance of 600 cubic feet per man; and, calculating the weight of an average Ayrshire cow at 8¼ cwt, and that of a man at 11 stone, the one will be exactly six times that of the other…If in a byre the air is to be maintained as pure as it is in our Army barracks, and the same provision is made in both cases for ventilation, then the cubic air space necessary for an Ayrshire cow, according to this basis, will be 600 ft. × 6 = 3,600 cubic feet. This is a cubic space six or seven times greater than the average cow is allowed, so that either the soldier gets too much space, or the cow too little, or else the basis of calculation is altogether wrong. Besides Army barracks, we have, however, other instances where the Government stipulate for a certain cubic space for each human being. A well known one is our Board Schools. In this case 10 square feet of floor space is demanded for each child…" (Speir 1896: 8)

> "The principal point to be attended to in keeping the animals clean and comfortable is to have the stall of a size suited to the size of the cow. This is attained by having stalls of different dimensions in different parts of the byre, and some farmers manage this by having one lot of stalls of one size and one lot of another; but in my own case I prefer to have them graduating from the largest to the smallest, i.e., the kerb or edge of the bed is, say, 8ft. from the front stall or partition at one end of a row of cows, and 7ft. at the other, so that the stalls graduate from one to the other. In this way the cows are sized out in regular order, and the perspective and general appearance of a byre is greatly improved when they stand in order, thus, like a regiment of soldiers." (McConnell 1899: 85)

This imposition of sanitary discipline was driven in part by economic exigencies, as its financial benefits gradually came to be more widely acknowledged, but it also possessed a dynamic of its own, bound up with the growth of technico-political power and the authority of experts. It was right at the generative nexus of "culture" and "nature", meticulously policing the ontological boundary between the two domains and in the process bringing it into being, not just discursively but materially, and with very material effects:

> "Since the passing of the Local Government Act of 1889, and the Public Health Acts, 1867, 1870, and 1878, with the Orders in Council relating thereto, sanitary matters pertaining to byres, stables, etc. have been enforced in a manner previously unknown." (Speir 1896: 7)

> "A few years ago there was no Act controlling the sanitation of dairies and cowsheds. Today an army of sanitary inspectors periodically visit the dairies and cow-sheds to ensure their being in a proper sanitary condition." (Pocock 1898: 19)

The principal political technological tool of this sanitary practice was an internal/external dichotomy, by means of which it sought to break down the animal body and its environment into discrete functional elements. This allowed an essentially Cartesian distinction to be made, at once ontological and strategic, between what was inside the animal and what was outside, facilitating a careful mapping of the various thresholds of the body which might carry the infectious bacteria from the animal interior to the human exterior, which is to say from "nature" to "culture". Thus the potentially infected animal body became a material problem of governmentality conceived in terms of pollution and containment–how to manage the body's processes such as to filter the spread of what it "contained" to the "outside":

> "It will be convenient if I contemplate the subject in two distinct aspects–milk, as susceptible of contamination from within, from the animal's own system; and then as a fluid liable to be contaminated by contact with certain noxious matters which exist outside the animal's body." (Brown 1887: 44)

> "Most of these disease germs come from outside contamination; but there is one class which we have every reason to believe is very commonly found in milk which comes from the cow itself, viz. the tubercle bacillus–that which produces tuberculosis or consumption." (Aikman 1896: 45)

> "There are various ways in which the bacilli bred in the interior of an animal may gain access to the outer world." (McFadyean 1895: 657)

This problem was posed most starkly in relation to dairying, which was then beginning a long period of steady expansion, for milk, after all, is easily read as the product of an animal's interior which is extracted in order to be consumed externally:

> "It is possible, but by no means proved, that the infective matter was conveyed through the agency of the milk." (Brown 1887: 52)

> "It is within the knowledge of all our readers that milk not infrequently acts as a propagator of disease by being the medium for carrying certain pathogenic or disease-producing germs." (Aikman 1896: 47)

> "The re-introduction of foot-and-mouth disease into our dairy herds adds another to the list of contagious disorders communicable to man through the medium of milk." (Wortley Axe 1892: 64)

> "There exists no doubt in the minds of the most advanced pathologists that human tuberculosis is largely due to specifically infected milk." (Wortley Axe 1892: 65)

Meat passes through a critical transformation before being consumed by humans; in the practice of cooking it is separated from the animal body, the traces of nature are effaced before consumption, and in the process it becomes not flesh but "meat", and part of "culture".[23] Milk on the other hand, which is necessarily extracted from a living animal, underwent as yet no such "civilising process":

> "Meat is always cooked before being used as human food, and it has been demonstrated by Koch, Lingard, and others that a boiling temperature for a few minutes is sufficient to destroy the vitality and infective power of the bacillus; but milk is generally used as food without cooking, more especially in the case of children. Tuberculous milk must therefore be looked upon as dangerous and likely to be the means of producing the disease in young or weakly subjects consuming it." (Duguid 1890: 314)

Now both the scale of the liquid milk trade and its importance within British dairy agriculture increased significantly in the last decades of the nineteenth century (Fussel 1966: 56). This was facilitated by the expansion of the railways and driven by the pressure of foreign competition in the less perishable dairy products brought to bear by new refrigeration technology. Against this background, tuberculosis–the zoonose most readily transmissible through milk–became the object for a network of disciplines embodying various specialised knowledges. These laboured to impose controls on the passage from internal to external, to purify it, render it amenable to human control, and subject it to the anthropocentric categories it was so apt to undermine; in this way they were integral to the technical realisation of humanist discourse at the heart of modernisation. This project was enormous in scope, as complex and fragmented as it was ambitious, and though well underway it was far from complete at the turn of the century (Duguid 1890: 51).

> "During the past eight to ten years tuberculosis, or consumption, has received more attention than any other disease of either man or animals. The reason for this has been not only that the belief in its contagious nature was becoming more generally accepted, but the discovery by experiment that the disease in man and animals was the same, and inter-communicable from one to the other." (*Ibid*: 305)

With this in mind the significance of the development of pasteurization and of the tuberculin testing of dairy herds in the last years of the century can be begin to be appreciated, as can the importance of the whole ensemble of emergent practices and technologies associated with the drive for "clean milk" in the early twentieth century. Through such practices the complex "mangle" of hybrid and continuous material relations between human and nonhumans animals was fashioned into a culture/nature dualism.[24] These included innovations in transportation and delivery designed to reduce the risk of contamination, leading to the earliest moves towards the doorstep delivery of bottled milk;[25] together with the intensification of longstanding efforts to produce an efficient milking machine as the cornerstone of a strategy of increasing mediation between milker and cow, a process which would eventually, decades later, see the first fully mechanized dairies in which both cow and milk were untouched by human hands. Thus hybrid relations which endlessly criss-crossed "nature" and "culture" were progressively reshaped and reclassified until they were brought into line with the conceptual architecture of humanism. By such processes modern "culture" and the category of the "human" intrinsic to it were made materially and historically actual.

Conclusion

> "Man is only a recent invention, a figure not yet two centuries old, a new wrinkle in our knowledge, and he will disappear again as soon as that knowledge has discovered a new form." (Foucault 1970: xxiii).

The three to four decades following 1866 in British dairy farming saw the development of a nexus of practices and knowledges in terms of which the animal was discursively positioned and ontologically classified in contradistinction to the human. Their effect was to purify the boundary between the human and the animal, crucially by rendering the animal body–and particularly liquid milk–"safe" for human consumption. This accomplished far more than its most obviously practical purpose–preventing the zoonotic communication of tuberculosis to humans–for in the process it contributed, in concert with innumerable similar mechanisms operating ceaselessly, to the constitution of the very sphere of "culture"; that powerful myth at the heart of modernity of a uniquely human realm of meaning, above and beyond "nature". By literally disinfecting the problematic relation between human and animal the anthropocentric polarity was made possible between "nature"–the animal body or raw material, and "culture"–its consumption by humans. Technical control at the level of material details was the means of ontological control at the level of metaphysical abstractions. Milk was thereby "civilised" and rendered "clean"; where there had been a deeply problematic animal substance and a bearer of zoonotic "agency", now there was a purified cultural product, ripe for the mass marketing of the twentieth century, and no longer so pregnant with questions concerning our real relations with the other animals. This was above all a material accomplishment requiring constant vigilance and ceaseless labour for its continuing plausibility; it had to appear at all times to be the natural order of things.

The development and dissemination of sanitary discourse and veterinary pathology was intimately connected to this practical and epistemological operationalisation of humanist discourse; only by regulating and controlling animal bodies and animal diseases with precision could their intrinsically problematic relationship to the category of the "human" at the heart of modernity be suppressed and managed. The category of "animals" is not self-evident but defined and constituted by an array of discourse and practice. Administrative and organisational rationalisation was critical in the general diffusion and co-ordination of such practices, pointing towards the intimate relationship between humanist discourse with its ontological separation of humans from nonhumans and culture from nature, and modern governmentality with its construction of a public sphere, its distinctive technologies of rule, its particular political ways of knowing. Thus the humanist practices underpinning the constitution of a sphere of "culture" were articulated within distinctly modern forms of social and political organisation, which were themselves, in turn, dependent for their rationale upon the historical category of the "human" thereby produced. The internal connection was the resulting notion of the human individual as a cultural subject, a free and rational agent separate from nature–the secular "soul" of modern politics.

Hence the discourse of humanism is deeply political; it is inseparable from the epistemological conditions of possibility of modern forms of sociality and power. Its effect is not directly to control human subjects, though it certainly facilitates such control, but more importantly to classify and thus produce them as "human". By organising the historically specific ways in which it is possible to know oneself as "human" vis-à-vis the knowable "animal", humanism produces that definitive category without which "culture" itself would be unthinkable.

Notes

[1] For a clear statement of this position see Foucault (1970: 127)

[2] For a fascinating analysis of some of the ways in which it is increasingly difficult not to regard many non-human animal practices as "cultural", see McGrew (2003: 419-439).

[3] Periodization is necessarily inexact and provisional, but I am in broad agreement with Foucault when he posits "two great discontinuities in the *episteme* of Western culture: the first inaugurates the Classical age (roughly halfway through the seventeenth century) and the second, at the beginning of the nineteenth century, marks the beginning of the modern age. The order on the basis of which we think today does not have the same mode of being as that of the Classical thinkers" (Foucault 1970: xxii). In British agriculture this shift did not begin to make itself felt on any scale until the latter half of the nineteenth century. See Pickstone (2000: 3).

[4] "One has to dispense with the constituent subject, to get rid of the subject itself, that's to say, to arrive at an analysis which can account for the constitution of the subject within a historical framework. And this is what I would call a genealogy, that is, a form of history which can account for the constitution of knowledges, discourses, domains of objects, etc, without having to make reference to a subject which is either transcendental in relation to the field of events or runs in its empty sameness throughout the course of history" (Foucault 1991a: 59).

[5] "The smallest AIDS virus takes you from sex to the unconscious, then to Africa, tissue cultures, DNA and San Francisco, but the analysts, thinkers, journalists and decision makers will slice the delicate network traced by the virus for you into tidy

compartments where you will find only science, only economy, only social phenomena, only local news, only sentiment, only sex." (Latour 1993: 2).

[6] See for example Sontag (1978). For a useful overview of medical anthropology see Helman (2000).

[7] The deep anxieties surrounding zoonotic disease were demonstrated in the recent media panics triggered by the outbreaks of avian and swine influenza.

[8] For a powerful account of the theological basis of modern secularism, see John Milbank (1990).

[9] For an account of Thomas Aquinas' anthropocentrism, see Linzey (1994: 13-14)). See also Ryder (2000: 35).

[10] This is typically overlooked by the ecology paradigm, which is deeply rooted in anthropocentric and homogeneous categories of "nature" and "society", despite its emphasis upon their inter-relations.

[11] For a critical account of the "Great Chain of Being", see Singer (1976: 159), cited in Linzey (1994: 30).

[12] All ontological discourse is essentially theological in its form, for it necessarily involves the separation of being and practice into distinct and transcendental domains in an *apriori* division of what is only *a-posteriori* divisible. More specifically though, the binary ontology of humanism is palpably a secular translation of what is still a theological conception of humanity.

[13] "The origin of these diseases not being traced, they were considered to have arisen spontaneously, and persons who entered upon the very precarious occupation of farming were told that they must be ready to submit to such diseases as a part of the ordinary casualties of their business" (Duckham 1893: 266).

[14] The development of agricultural statistics then, through giving birth to the concept of animal "population", simultaneously gave rise to what Latour calls "the indisputable conflict between health and wealth", albeit the health of nonhumans. See Latour (1988: 16).

[15] "There is no difficulty in realising the fact that an animal suffering from infectious disease is capable of communicating the disease by the agency of the tissues and fluids of the body" (Brown 1887: 47).

[16] The analysis here parallels that of Bruno Latour in *The Pasteurization of France* (Latour 1988).

[17] Emphasising this attitudinal shift, G. T. Brown remarked that "Had the Privy Council of five-and-forty years ago attempted to arrest the progress of pleuro-pneumonia and foot-and-mouth disease by slaughter and stopping the free movement of animals, they would have raised a storm of opposition which would have rendered the effort futile" (Brown 1893: 283).

[18] "A tuberculous animal is a sort of hotbed for the propagation of the tubercle bacillus, and wherever such an animal exists it is a source of danger to other animals by reason of the bacilli which may be voided from its body" (McFadyean 1895: 657).

[19] "Cattle-Plague was introduced in 1865, and began to spread extensively owing to the absence of any restrictions on the movement of diseased or affected animals. Some time was spent in making inquiries as to the nature of the disease, and an Order was passed calling upon owners of diseased cattle to give notice of the

existence of disease; this, however, they neglected to do, preferring to send their stock to market as soon as a case of cattle plague was detected" (Brown 1893: 283).

[20] In reference to the proceedings of a Commons Select Committee on cattle diseases in 1873 to which he gave evidence, T. Duckham reported that "Evidence of much weight, both by agriculturalists and professional witnesses, was given, tending to show that such enactments as would be required to exterminate foot-and-mouth disease would meet such strong opposition that they would be difficult, if not impossible, to carry out." The Committee therefore recommended "that the Privy Council should cease to issue orders for the check of this disease" (Duckham 1893: 273).

[21] "To understand how animals were alive was to understand the relations of structural and functional relations between organs, *and*, additionally and crucially, the functional interactions between each animal system and its *milieu*. This is the origin of our meaning of "environment" as more than just 'surroundings'" (Pickstone 2000: 116).

[22] The Dairies and Cowsheds Orders of 1885 and 1899 required a certain minimum standard of accommodation, with specifications covering air space, light, ventilation and drainage. (Middleton 1909: 51).

[23] This is what is referred to as "the structure of the absent referent" (Adams 1996: 40, 42, 44, 47). Anthropologist Noëlie Vialles documents ethnographically the way that meat is "de-animalised" i.e. "separated from everything too obviously reminiscent of the animals itself" in French abattoirs (Vialles 1994: 127).

[24] So pervasive were these technologies that one dairy farmer, writing in the Journal of the British Dairy Farmers' Association, felt moved to complain that "One might almost conclude that milk has had enough attention paid to it by "experts" as from the cow to the consumer (and before and after) it is watched with solicitude; it is filtered, cooled, heated, bottled, sterilised, humanised, municipalised, and otherwise so maltreated that I am sure no self-respecting cow would own it; and an ordinary dairy farmer may be forgiven for enquiring into the necessity for all this." (Marshall 1906: 17).

[25] The first attempt to bottle milk was made by George Barham of the Express Dairy in 1884; the bottle was sealed with a wired-on cap which did not work in practice (Jenkins 1970).

Bibliography

Adams, C. 1996 *The Sexual Politics of Meat: A Feminist-Vegetarian Critical Theory*. New York: Continuum.

Aikman, C. M. 1896. "Pasteurized Milk". *Journal of the BDFA*. XI, Part I.

Berger, J. 1980 *About Cooking*. New York: Pantheon.

Brown, G. T. 1893. "The Progress of Legislation against Contagious Diseases of Livestock: Part II". *Journal of the RASE.* Third Series. IV.

—. 1887. "Milk in Relation to Infectious Diseases". *Journal of the BDFA*. III, Part I.

Duckham, T. 1893. "The Progress of Legislation against Contagious Diseases of Livestock: Part I". *Journal of the RASE.* Third Series, Volume IV.

Duguid, W. 1890. "Tuberculosis in Animals and its Relation to Consumption in Man" *Journal of the RASE.* Third Series. I.

Foucault, M. 1991a, "Truth and Power". In P. Rabinow (ed.), *The Foucault Reader*. London: Penguin: 51-75.

—. 1991b. *Discipline and Punish: The Birth of the Prison.* London: Penguin.

—. 1970. *The Order of Things: An Archaeology of the Human Sciences.* London: Tavistock.

Fussell, G. E. 1966. *The English Dairy Farmer 1500-1900.* London: Frank Cass & Co.

Helman, C. 2000 *Culture, Health and Illness.* Oxford: Butterworth-Heinemann.

Jenkins, A. 1970 *Drinka Pinta: The Story of Milk and the Industry that Serves it*. London: Heinemann.

Latour, B. 1993 *We Have Never Been Modern*. Harlow: Pearson.

—. 1988 *The Pasteurization of France*. Cambridge: Harvard University Press.

Linzey, A. 1994 *Animal Theology*. London: SCM.

McConnell, P. 1899 "The Best Form of Cow Stall". *Journal of the BDFA.* XIV, Part II.

McFadyean, J. 1895 "The Prevention of Tuberculosis in Cattle". *Journal of the RASE.* Third Series. VI.

McGrew, W. C. 2003 "Ten Dispatches From the Chimpanzee Culture Wars". In F. de Waal and P. Tyack (eds.), *Animal Social Complexity: Intelligence, Culture, and Individualised Societies.* Cambridge: Harvard University Press: 419-439.

Marshall, C. 1906 "The Desirability or Otherwise of the Municipal Supply of Milk–From the Dairy Farmer's Point of View". *Journal of the BDFA*. XX.

Michael, M. 1996 *Constructing Identities: The Social, the Nonhuman and Change*. London: Sage.

Middleton, C. 1909 "Suggested Amendments of the Dairies, Cowsheds, and Milkshops Orders in Relation to Buildings and Inspection". *Journal of the BDFA*. XXIII.

Milbank, J. 1990 *Theology and Social Theory: Beyond Secular Reason.* Oxford: Blackwell.

Pickstone, J. 2000 *Ways of Knowing: A New History of Science, Technology and Medicine.* Manchester: Manchester University Press.

Pocock, S. 1898 "Tuberculosis of Cattle–From the Dairyman's Point of View". *Journal of the BDFA*. XIII, Part II.

Ryder, R. 2000 *Animal Revolution: Changing Attitudes towards Speciesism.* Oxford: Berg.

Singer, P. 1976 "All Animals are Equal". In T. Regan and P. Singer (eds.) *Animal Rights and Human Obligations.* New Jersey: Prentice Hall: 148-162.

Sontag, S. 1978 *Illness as Metaphor*. London:Vintage Books.

Speir, J. 1896 "The Air of Cow Houses". *Journal of the BDFA*. XI, Part I.

Vialles, Noëlie 1994 *Animal to Edible*. Cambridge: Cambridge University Press.

Wortley Axe, J. 1892 "Milk and Disease". *Journal of the BDFA*. II.

—. 1896 "Sources of Milk Infection". *Journal of the BDFA*. XI, Part I

CHAPTER THREE

ANIMALS AND THEIR POSSESSIONS: PROPERTIES OF HERD ANIMALS IN THE ANDES AND EUROPE

PENELOPE DRANSART

Since the nineteenth century, social anthropologists have conceived of the ownership of property as property relations, in the sense that the holding of property should be understood in the context of "social relations between people" (Hann 1998: 4). The notion that human beings may claim quadrupeds as private property constitutes one of the assumptions underlying anthropological literature dealing with pastoralism, in that herd animals belong to people in a web of social relations between the human owners (Ingold 1980; Dransart 1996). Dahl and Hjort (1976) succinctly summed up the proprietorial assumption of human beings as owners of herd animals in the title of their book *Having Herds*.

Herd animals in pastoralist societies are not seen as exercising control over the products of their body or their labour. Therefore anthropologists generally do not perceive them to be "acting subjects" (to borrow a phrase from Strathern 1984: 162) and they tend to conceive of animals as alienable property. Strathern used the phrase "acting subjects" in the context of a discussion of men's claims to own land and their recognition of women as "wealth" in the Mount Hagen area of Papua New Guinea. She argued that Hageners do not believe women to be objects, because Hageners do not oppose things to persons "as our own subject-object matrix postulates" (Strathern 1984: 165). In a later work, Strathern (1988: 176) developed her analysis of Hagen notions of personhood through what she calls "the symbolic mechanisms or techniques" of reification and personification. The status of herd animals seems a more intractable problem than that of women or even slaves as far as the human recognition of animals exercising control over their properties is concerned. As Ingold

(1994: 16) stated, herd animals "are cared for, but they are not themselves empowered to care".

In introducing my theme, I appeal to studies concerning concepts of personification among human beings as a contribution towards the development of a fuller anthropological understanding of the inter-agentive relations between human beings and the animals they herd. It is true that quadrupeds are incorporated into social relations between human owners, but they are also beings that participate in their own social organisation. Tim Ingold (1994) proposed the term "inter-agentivity" in a discussion on human perceptions of social systems in which wild animals and human beings "attend" to each other. From my experience of having been a herder during extensive periods of participant observation in a particular social setting in which human herders care for their herd animals on a daily basis, I have been prompted to wonder whether animals herded by human owners might be able to reciprocate and whether their owners might, indeed, regard them as being the possessors of possessions.

Possessions: Blood, Fleece, Names and Leadership

As part of this enquiry, it is necessary to investigate the notion of what being an "animal" means in two different (Andean and European) cultural contexts before considering how organisms with circulatory systems have been categorized. The distinction is one of "herders" and "herded". The notion of animal being will be explored through various tangible and intangible entities. This discussion considers items with material substance (blood and hair), qualities (names assigned to categories) and roles (leadership). Whereas the consumption of blood and its use in transfusions between species and amongst conspecifics reveals cross-cultural differences, there is consensus over the value of hair/fleece as a material for clothing the human body in the different ethnographic contexts examined here.

In the anthropological literature, blood and hair have been termed exuviae and they both exemplify what Gell (1998: 112) called an "exuvial substance …index of life, growth and reproductivity". While the wearing by human beings of clothing made from the fleece (hair) of camelids does not seem to provoke feelings of unease or distaste in either of the ethnographic contexts discussed here, the use of blood is complex. Aymara speakers are often reluctant to accept blood transfusions from non-kin and they do not bleed their camelids as a source of food.[1] Some of the complexities towards blood and the notion of animated beings in European attitudes, on the one hand, and in Andean attitudes, on the other, are explored below.

Nomenclature deserves attention because the Aymara language, which is spoken by the bilingual herders among whom I have done fieldwork in Isluga in the far north of Chile, does not have a word equivalent to "animal" in the sense of an inclusive term as is available in European languages. My particular concern in this paper is to explore the inalienable properties of llamas and alpacas (including their social behaviour and leadership roles) in the context of the domesticating activities of their human owners. South American camelids have been imported into Britain since the nineteenth century (Ritvo 1987: 234, 239), but it is only recently that growing numbers of smallholders have begun to keep them. The different strategies adopted by Aymara and British herders of camelids are instructive, because they demonstrate different responses to the properties of herd animals in the interrelationship between the respective social relations of human beings and camelids. These differences raise questions relating to animals as possessors of possessions, as alluded to above. Since there may be a tendency for possessors to think of themselves as the named owner of their possessions, this discussion will tackle some particularly salient aspects concerning the naming of animals. Differences in categorization are associated with different understandings of caring or herding practices in the two case studies examined in this chapter. These naming practices are related to local understandings of physiological knowledge.

If, as Franklin (2001: 3) has argued, sheep in Britain constitute "an animal category dense with paradoxical significance" and sheepwatching has become a means for observing transformations in local-global connections, then it must be recognised that camelids have played an analogous and equally central role in a deeply layered politicised process of interconnections between South America and the rest of the world. In the whole of the Americas, camelids were the only ungulates to be subject to the caring (domesticating) attention of human beings before the introduction of European herd animals from the sixteenth century.

In the Andes, camelids have served for millennia as labourers, gifts and commodities in a network of relationships among and between communities of human and herd animals.[2] I consider the recognition of leadership qualities by the owners of camelids as one of a series of possessions, whether intangible or tangible. My consideration of such possessions is located in a discussion of what an animate being is in historically situated European and Andean contexts. The focus here is on how people have understood the physiological processes of living beings and how such understandings have been put to practical use in herding camelids.

Being "Animal" in a European Cultural Matrix

In the English language, the term "animal" incorporates human beings as a category (Willis 1990: 19). An animal is an organised being possessing life, sensation and the capacity for voluntary movement. Harriet Ritvo (1987: 12) noted that eighteenth-century formulations of taxonomies treated the term as an inclusive category. She cited the definition in the first Encyclopaedia Britannica (published in 1771) that an animal is "an organized body endowed with sensation". The term has its etymological root in the Latin *anima* (air, breath or life); therefore it is connected with the notion of "soul". Certain seventeenth-century writers, such as Descartes, Malebranche and Regis, regarded what were called "animal spirits" as a property of the bloodstream. In so doing they recognized the sentience of animals but at the same time they denied animals rationality as possessors of souls. Such views appear to have antecedents in beliefs treated in ancient Greek texts, which refer to the spirits of the dead craving "dark-clouding blood" of sacrificial animals to enable them to recall memories of life on earth (Homer Bk XI: 36, 152), and from a Roman understanding of steam emanating from warm human blood to be the soul escaping from the body (Cumont 1959 [1922]: 51).

Seventeenth-century philosophers viewed matter as obeying laws of motion; they thought that mechanics provided a means of explaining material change. However there were differences of interpretation between them. Whereas Descartes claimed to know the essence of material form, Locke did not (Ayers 1981: 250). Locke's dual definition of "Man" drew attention to substance as possessing a corpuscular structure that was subject to his idea of mechanics. Such a being, Locke said, should first be regarded as possessing "solid extended Substance, having Life, Sense, spontaneous Motion, and the Faculty of Reasoning". Second it is "a rational *Animal*; which by the several definitions of *Animals*, *Vivens*, and *Corpus*, resolves it self into those enumerated Ideas".[3] According to Yolton (2001: 682-3 and 675), Locke argued that to become a rational being, one had to learn to control desires and passions and, in so doing, one became the proprietor of one's own person. Human beings were, therefore, seen as distinct from animals through the possession of rationality although corpuscular particles were understood to course through both animal and human bodies.

Against the prevailing climates of belief, European doctors began experimenting with intravenous infusions of blood, first on dogs. The first person credited with doing this successfully in England was Dr, later Sir, Christopher Wren (Farr 1980: 143-4). In 1667-68 the French doctor Jean

Denis transfused lamb's blood into human patients (Brown 1948; Farr 1980). The rationale to which Denis appealed was based on an analogy between transfusion and feeding. He recognised that the blood of different animals "must be poyson and venom in respect of [one] another" but that "the strength and goodness of the meats and drinks we take, is able to correct the ill-temperament of the blood and render it better" (Denis, cited by Farr 1980: 150). Denis did not take blood from a donor of the same species as the patient – he thought that such a strategy would pose an unacceptable risk to the donor (Farr 1980: 151). The donors and patients of Denis survived, but Farr (1980: 162) questioned how much blood Denis took from the donor to give to his patients – it might have been comparatively little. By 1825 doctors generally recognised that blood from different species is incompatible for use in intravenous infusion. The analogy between the taking of "living" blood (rather than extracting it from a slaughtered carcase) and "food" is difficult to conceive of in an Andean context where herd animals are not bled as a source of food while they are still alive.

Andean herders do not milk llamas or alpacas as a source of food, either, and there are no indications that they did so in the past (Dransart 2002: 12). Nevertheless, it seems that long-standing ideas pairing milk and blood have featured in both the conceptual environments considered in this chapter. In Europe, Isidore of Seville appealed to ancient Greek etymology to claim that "after the birth whatever blood has not yet been spent in the nourishing of the womb flows by a natural passage to the breasts, and whitening by their virtue, receives the quality of milk" (*Etymologiæ* 11.1.77, cited by Wood 1981: 719). This view was repeated by later medieval commentators (McLaughlin 1974: 115). In the Andes, there is an absence of documentary evidence and the breast often serves metonymically to denote milk. Aymara-Spanish dictionaries commonly list *ñuñu* (breast, nipple or teat) and its verb *ñuñuna* (to suckle) (Bertonio 1984 [1612]: 288; Apaza Suca *et al.* 1984: 147; Mamani Mamani 2002: 114). Yet there are also a few references to *wilawi* (Apaza Suca *et al.* 1984: 250) and *wilali*, a "vital liquid", as terms used for breast milk (Szemiński 1987: 17). The root of these terms is *wila*, the Aymara for blood, and the parallel quality of blood and milk seems to be recognised in a saying that in the womb the placenta suckles the foetus with maternal blood.[4]

In Europe the development of categories took a paradoxical turn in reference to milk.[5] During the course of the eighteenth century, Linnaeus proposed, developed and refined the taxonomy and nomenclature of animals. His work, entitled *Systema naturae*, grew greatly in page numbers as it went through different editions until the last (twelfth), which Linnaeus prepared in 1766. Londa Schiebinger (1993: 384) regarded the tenth edition as "epoch-making" because in it Linnaeus presented binominal names (i.e. generic and specific) of nearly 4,400 animal species. However, she points out that the classificatory work of Linnaeus emerged from a tradition that can be traced back to Aristotle.

In the Aristotelian tradition, organisms were understood to belong to a great chain divided into more or less arbitrary categories (Encyclopaedia Britannica 1977: 766-7). The term Linnaeus coined for the group previously known as quadrupedia, Mammalia ("of the breast"), was derived from mammae which, as Schiebinger (1993: 388) noted, were not a universal feature of the class in question. Given the different connotations connected with the function of breasts in European cultural history, she argues lactation was both "less than human" and "more than human" (Schiebinger 1993: 398).

She observed that Linnaeus unintentionally devised a gender-charged term for Mammals at a time when authors, including Linnaeus, wrote tracts to urge mothers to cease the practice of having their babies wet-nursed. The "complex cultural matrices" to which Schiebinger (1993: 411) referred still resonate today. Later I will discuss how British camelid herders use a gender-charged terminology for their animals in a technology of domination over their animals.

Being in an Andean Cultural Matrix

The Aymara language does not have a term to encompass human and non-human animals. In an early seventeenth-century grammar, Torres Rubio (1616: 24) observed that Aymara did not have abstract nouns and that to convey such ideas a speaker had to use a verbal form of the concrete noun or duplicate it. The precision with which Aymara speakers use their language means that terms for certain super-ordinate categories are not necessary. Human beings and herd animals are not held to have properties in common that would lead to them being named under the same category. Whereas "people" are *jaqi*, Aymara-Spanish dictionaries provide the Spanish "*animal* (*doméstico*)" for the term "*uywa*", which literally indicates a creature cared for by a human owner (for an example from a dictionary of northern Chilean Aymara-Spanish, see Mamani Mamani 2002: 188).

An early seventeenth century dictionary compiled by Ludovico Bertonio (1984 [1612]: 53) explained the lack of a term for "animal" in the following manner: "There is no generic name which comprehends all:

those with four feet they call Pusicayuni".[6] Gonçalez Holguín (1952 [1608]: 208) listed an equivalent term in Quechua: "tahuachaquiyoccuna is taken for all animals, or those which are not human beings".[7] Bertonio's term *pusicayuni* and Gonçalez Holguín's *tahuachaquiyoccuna*, both of which are literal renditions of "four-footed" or "four-legged", might have been late sixteenth to early seventeenth century neologisms that did not gain currency in the Andes. It should be noted that such terminology is consonant with the European category of quadrupedia used in an Aristotelian tradition. On the basis of working with native bilingual speakers of Aymara and Spanish, Lucy Therina Briggs (1979: 88) commented that the publications of Bertonio and Torres Rubio contain literal translations from Spanish to Aymara that today sound awkward to a native Aymara speaker.

The term *uywa*, as used in Isluga in certain contexts, implies a contrast with undomesticated animals, which are animals cared for by the spirits of the hills. I have previously argued that "domestication" in Isluga, northern Chile, is the product of the daily caring activities that herders undertake and that herders bring each generation of herd animal into a state of domestication (Dransart 1996; 2002: 17, 47). Herders in Isluga work with the social properties of the herd in a daily and yearly round that recognises co-operative collaboration of human and herd animal practices which are conducted in a terrain considered (by the herders) to have its own animated characteristics (Dransart 2002: 80-1).

Therefore, the "being" of humans or *uywa* must be understood as a category that functions in a cultural matrix that differs radically from European matrices. Claudette Kemper Columbus (2006: 122) refers to a

> "complex system of associations in an animist cosmos–by which, in this context, the term does not refer to everything having a soul, but to everything being animated and alive in a social ambience not exclusively human and where key terms reveal the internal framework of thought and relationships among human and non-human actors".

In my understanding of the matter, Isluga concepts of animation are intimately related to inhaling and exhaling air. Weaving constitutes an analogy for such processes, and the exuviae of llamas and alpacas provides the raw material. Isluga weavers term the outermost threads of the warp, which are held taut by the loom, the "mouth" (*laka*), which opens and closes to receive the weft each time the weaver changes the shed.[8] In striped textiles used to make bags, the central stripe is termed *chuyma*, metaphorically the "heart" or the centre of the weaving, but more literally its lungs, which breathe in the air to circulate through the web (Dransart 2002: 122-3).

Isluga concepts of the soul relate to the notion of exhaling and halitosis in the escape of the soul from the body at death and also to beliefs that the souls of the dead sustain the fertility of the world in which people and camelids live (Ortega Perrier 2001). In her analysis, Marietta Ortega Perrier relates the cycling of blood/water through bodies and the land to metaphors of the soul. This theme has also been addressed by Joseph Bastien (1985) in a discussion of Qollahuaya body concepts in Bolivia.[9] The dance of flute players in an agricultural ritual called Khallay Chajmay involved the participants in a spiral, anticlockwise movement from east to west followed by an outward movement from west to east. Bastien (1985: 598) perceived this dance to be an analogy of the circulatory, respiratory, digestive and reproductive processes in a living body with "its centripetal and centrifugal forces of circulation" of liquids. In this understanding of physiology, air, blood and fat are the most important fluids and the heart is the "distillation center":

> "Air is the invisible fluid substance that provides breath to living things. People, plants, and animals share in this fluid, which is understood as a unitive principle between them." (Bastien 1985: 599).

That the dancers played wind instruments presumably is significant in his discussion of the animating properties of air.[10] In this context, it is worth recalling an account by Verónica Cereceda (1993) of Chipaya spatial and temporal concepts of the wind, whistling round the Bolivian highlands, through the Chilean valleys traversed by Isluga people when they leave the altiplano for the coast, and back to the highlands. In integrating and unifying human and non-human actors, the wind is accorded a leadership role, which is a quality Isluga herders also expect their guide llamas and alpacas to demonstrate (Figure 3-1).

Although herders in Isluga exercise a hierarchical control over the fate of the individual animals in their herd, a fundamental characteristic of the bond of possession is tempered by a notion of parallelism. Families have a special relationship with specific geographical features in Isluga territory. People recognize certain places as *uywiri*, meaning "herder" or perhaps "creator". An important category of *uywiri* is known by the term *juturi*, which are deep wet or dry hollows associated with the inner world, considered to be the source of origin of llamas, alpacas and sheep (Dransart 2002: 59-60). Hence the notion of human ownership of camelids is mediated by the *uywiri*, since human herders care for the animals provided through the beneficence of their *uywiri*. Human origins are more

Figure 3-1: Female guide llama (*tilantir*) leading the herd into a corral for the marking ceremony, Isluga, Chile, 1987 (photograph: P. Dransart).

complex; the sources of a previous world generation of gentiles are still visible in remote locations termed *thaqsu*, which are described as being corral-like but are also said never to have functioned as enclosures for camelids or sheep (Dransart 2002: 61). The point here is that both human beings and camelids/sheep are, in a sense, owned by the community's *uywiri* but people and camelids have different points of origin in Isluga territory.

Another factor which points to the parallel character of the relationship between people and herd animals is that Isluga herders judge camelid behaviour and leadership on grounds that are appropriate to camelid behaviour rather than that of surrogate slaves or children. There is a clear contrast between their herding practices, which allow for autonomy in the behaviour of their camelids, especially in the crucial role of leadership, and those of European and North American smallholders, who assume the role of leadership and run the risks of having young camelids imprint upon people.

Having and Holding

Since at least the nineteenth century, people have attempted to raise camelids in non-Andean countries.[11] Gramiccia (1988: 22) reported that D. Burnett Esq. managed to raise fifteen young alpacas in Berkshire from 1817 onwards. However, throughout the nineteenth century, owners seemed to have had limited success in enabling the herd to reproduce in non-Andean countries. A notable example of camelid transportation is provided by Charles Ledger, who exported a herd from Bolivia to New South Wales, Australia, arriving in 1858, where they were mortgaged on account of their owner's debts (Gramiccia 1988: 76-7). Through the refusal of the camelids to reproduce, it can be argued that they eluded the full domination of their human owners.

It is only relatively recently that small herds of camelids have become visible in the countryside in Britain.[12] The British Camelids Association was founded in 1987 with the expressed aim "to represent everyone interested in llamas, alpacas, guanacos, vicunas [sic] and camels, be they owners or not" (2003/2004 Handbook). At the time the Association was coming into existence, I was already doing fieldwork in the far north of Chile. Pastoralists in the highlands east of Arica, an area which is adjacent to, and within easier reach of larger exporting centres of camelid fibre than Isluga, voiced their concerns to me in 1986 that if farmers in non-Andean countries were to herd alpacas for fibre production, their livelihood would suffer. To add to their worries, they thought that the competition would be made more unequal because camelids in other countries would have access to better pasture lands than in the Andes. The vegetation in herding grounds at altitudes of over 4,000 metres above sea level is steppe-like and, apart from areas of artificially irrigated moist pasture, sparse. However, their worst concerns are seemingly not being realised. British farmers have not adopted camelids as a means of circumventing EU quotas on other herd animals and, although there are a few substantially-sized herds, it is mainly smallholders who are acquiring them.

British smallholders purchase camelids for various reasons. The Association's handbook (2003/04: 10) lists the following: companion animal, fibre (specifically mentioning alpacas), packing and trekking, and guarding sheep from foxes or stray dogs.[13] What I wish to do here is to examine the nomenclature and training practices that smallholders use as part of their herding strategies as a means of domesticating camelids. British camelid owners, like their counterparts in other countries such as the U.S.A., have devised their own domesticating strategies (see, for example, Burt 1991). These strategies are significantly different from those adopted by Andean herders. Here, I will examine three such strategies of domestication.

First, language constitutes a form of control. Ritvo (1987: 51) has indicated that human beings have the power to restructure and re-create reality in their dealings with non-human species through discourse: "[a]nimals never talk back". European and North American owners are renaming Andean categories for camelids. At times it seems arbitrary; as an example, English-speaking smallholders have adopted the generic Spanish term for all young animals, *cría*, to designate young camelids. On other occasions the renaming is gender-charged: the term "girls" is applied to female alpacas and "sires" to the stud males they visit for mating purposes. Such nomenclature is related to forms of language that are not politically neutral, as Schiebinger (cited above) detected in her study of Linnaean categories.

Second, the British Camelids Association wishes to have guanacos removed from the "wild" category. This aspiration constitutes a clear desire to exercise power relations over an animal which the Association has declared to be tractable. Such a perception of guanaco behaviour is modelled on deportment that llamas and alpacas have been taught to display in non-Andean countries. The human owners of camelids have certain expectations of their behaviour that revolves round the notion of appropriate handling.

Animals which do not conform to the range of behaviours perceived to be normal may be labelled as deviant under the category of "berserk male syndrome" or "aberrant behaviour syndrome" (Gegner 2000: 9; Stamberg and Wilson 2007). The syndrome is attributed to camelids, especially males, who are given too much attention and handling by their owners before they are weaned. Stamberg and Wilson (2007) regard such a training as a form of abnormal socialization towards human beings, with the result that the camelid potentially becomes aggressive. The threatening behaviour displayed includes chest butting, biting and attempting to wrestle with the legs of people. A number of internet sites offer training advice to smallholders. Instructions typically include limiting the handling of camelids and preventing them from invading "your personal space" (Stamberg and Wilson 2007).

The language in which this advice is couched is of interest, given the gender terminology mentioned under point one, above. An example is provided by a website provided by John Mallon, who offers camelid training clinics:

> "[I]t's time to talk about one of the most unpleasant topics relating to llamas and alpacas – Berserk Male Syndrome, or BMS. Actually, in these

politically correct times, the new term is ABS, or Aberrant Behaviour Syndrome, so as not to leave out the "ladies" (and, yes, females do develop the behaviour, although usually to a lesser degree, as they are not typically the territorial defenders that males are.) …. [T]o the best of my knowledge, no one has ever deliberately created a berserker … yet there are more and more of them out there – how can this be? If you've never encountered an ABS animal, you'll find it difficult to imagine one. It is an animal which will scream, spit, charge, attack, bite, butt and lay on top of people – it is a llama or alpaca that can kill….[M]ost of the animals that are labelled "berserk" are not – they are simply spoiled, disrespectful, bad-mannered, undisciplined "brats" that have been "trained" in an overly permissive manner and bribed with hand-fed food treats….Its [sic] simply a matter of understanding the psychology of the prey animal, and communicating to him in a way that he understands that he is not allowed to push people around. Once he understands that, he easily accepts it and is unlikely to regress, unless, of course, his handler goes back to his old habits. (That's the beauty of the herd animal's makeup – his willingness to follow leadership unquestioningly …)." (Mallon 1998-2002).

There are other web sites which also warn owners against too much physical contact with their camelids, especially when bottle feeding them. Advice is also given against raising camelids on their own without the presence of a herd. Mallon attributes problematic behaviour in a male animal to the onset of puberty when

> "[H]ormones start to run his life (remember being, or raising, a teenager?) The male, being the designated territorial defender, then attacks his human handlers when they enter his paddock (territory) to drive them out, just as he would any other llama or alpaca on "his turf"." (Mallon 1998-2002).

The categorization at work in his web site is that the super-ordinate term "animal" includes humans and camelids (both experience the onset of puberty), but that the former belong to the sub-category of "predators" and the latter to "prey". Mallon (1998-2002) reminds his readers that "what is submissive behaviour in a predator (initiating physical contact, for instance) is aggressive behaviour in a herd-living prey species". The advice he provides to human owners is that they should learn to handle young animals in an appropriate manner, which he terms "businesslike", to resist the urge to cosset and to train the animals to accept a certain degree of handling.

The "holding" metaphor of Western smallholding practices is apposite because, even with the use of an appropriate level of handling, the camelids are subject to far more human touch than is the case with their Andean counterparts. This handling includes grooming. Toenail trimming is an operation that has to be performed and camelids must be taught to tolerate having their feet lifted. In the Andes it is not necessary because the nails are worn down by the rocky, sandy terrain and the long distances that camelids walk on a daily basis from the sparse shrub vegetation high in the hills to the wet, low-lying pastures. Brushing the fleece is another practice to which camelids are subjected in European and North American countries. Andean camelids wallow in sand hollows instead and do not require regular human intervention.

Third, in their herding strategies, smallholders use halters which have been specifically designed to accommodate the structure of the camelid skull, especially in alpacas (Bennett 2002).[14] The halter is a device that enables the human owner to lead. It does not enable camelids to act as their own leaders within the social unit of the herd. Many owners in Britain have so few camelids that they do not encourage the development of leadership qualities within their herd. [15] Instead they train their camelids to wear halters so that they can take their animals to go where they want to take them, for example, to shows (Figure 3.2). Although the instructions in the published literature and websites indicates the relative ease involved in training a camelid to wear a halter, the frequent putting on and removal of a halter adds to the handling mentioned above. Owners are warned not to leave their camelids wearing a halter for long periods of time, or to attach the lead to a tree or fence post because it is possible for a llama or an alpaca to break its neck by attempting to break free from a tether without an elastic extension (Gegner 2000: 3).

The Isluga herding strategies I observed (and in which I participated) provide scope for recognising the autonomy of llamas and alpacas, especially in allowing the camelids to develop leadership from within the herd. Hence camelids in Isluga act as subjects within the herd and, in that context, they own their autonomy. A characteristic of herding in Isluga is that llamas understand two verbal commands: *kuti* ("turn round") and *piska* ("keep going") (Dransart 2002: 65). As soon as the guide animal of the herd (known as *tilantir*, a term borrowed from the Spanish for "leader"),[16] hears the word *kuti*, she will turn. The rest of the herd will, one by one, turn to follow her. What I find particularly striking is that human herders never train young llamas to obey these verbal commands. Instead the young animals learn from their elders among their own species.

In some parts of the Andes herds of llamas and alpacas are left to follow their daily routine largely unsupervised by their human owners. Isluga practice is for a family member to herd their animals for as much of the day as is practicable. However, Isluga people know that, if left to their own devices, the guide animals will take the herd down to the moist pastures

Figure 3-2: Alpaca wearing a halter at the Royal Welsh Smallholder and Garden Festival, Builth Wells, Wales, 2005 (photograph: P. Dransart).

and return uphill to the steppe-like shrub and grass lands in the evening without human instruction.

Therefore, in contrast to the notion of parallelism with which I characterised human-herd animal relationships in Isluga, which recognises the autonomy of species, I would apply the notion of parallax to the smallholders and their camelids.[17] By parallax, I am referring to an obtuse relationship between human and herd animals in which the human beings interact with their herd animals in a hierarchical order. An illustration of such hierarchical control is provided by Tani (1996), who argued that, in the Mediterranean and Middle East, herd animals have for millennia acted like serfs in a household economy in which the head (*dominus*) dominated his movable living property, both quadripedal and bipedal. An ethnographic detail exemplified his argument; in Abruzzo, central Italy, a shepherd castrates a ram and trains him by means of a rope round the neck to serve as a bell-wether. This example is instructive because Tani (1996: 389) commented that the shepherd gives the bell-wether a name such as "Generale" or "Mussolini".

Concluding Comments

In the context of the cultural trajectory that characterises many Western practices, it is possible to envisage how seventeenth-century doctors could imagine conducting experiments by transfusing blood from a donor lamb into the veins of a human recipient and how Welsh smallholders might have and hold their camelids at the end of a lead attached to a halter. On the other hand, the Isluga practice of scattering the blood of a slaughtered llama or alpaca to the *uywiri* before people can partake of the sacrificed blood and meat of the dead animal acquires importance in the transactions that take place between territorially based human and non-human herders (*uywiri*). The concept of caring for herd animals takes place in a context of land tenure that is significantly different from that of western Europe. In Isluga, territory is criss-crossed with rights of access and systems of social obligations. Areas of pasture are unfenced. People refer to European-style fields by the Spanish term "*parcelas*", which indicates an enclosed portion of land. The fenced "paddock" to which John Mallon referred is symptomatic of a series of constraints which Western camelid owners impose on the development of social leadership among their small herds.

There is something about Isluga notions of camelid ownership that is evoked by a statement made by Strathern (1984: 170) in reference to the personhood of Melanesian women: llamas and alpacas are "both part of and detachable from" the body of the *marka*, which is constituted by the Isluga social collectivity of people and other species. In the further development of her theme, Strathern (1988: 190) observed that social forms "are revealed through distinctions, while the 'construction' of the distinctions themselves, the classificatory scheme they constitute, the symbolic relations as we would have it, remain implicit".

The constructions of two different classificatory schemes have been considered in this chapter with reference to the physiological processes of living beings. My discussion of socially mediated strategies of Welsh smallholders has pointed to a hierarchy of domination which masks the recognition (among the human owners) that herd animals are capable of exercising control over their own properties. This study of complex cultural histories concerning attitudes towards animate beings endowed with sensation in Andean and Western conceptualisations reveals important differences. These disparities have impacted on current social practices regarding animals and property relations and have resulted in herders adopting dissimilar herding strategies in the two geographical regions.

Acknowledgments

I owe the people of Isluga, northern Chile, a debt of gratitude for making my fieldwork possible since 1986. In addition I wish to thank the smallholders who attended the Royal Welsh Smallholder and Garden Festival, Builth Wells, Wales, 2005 and other smallholders in Wales who responded to my questions on the research reported here. Funding for my research is listed fully in Dransart (2002: xvi), to which I wish to add an acknowledgement for further financial support from the Pantyfedwn Fund of the University of Wales, Lampeter in 2003. The writing of this paper was finalised in 2007, when I was Visiting Fellow at the Sainsbury Research Unit, University of East Anglia, and I wish to thank staff at the Unit for the support they provided me. Finally, I wish to thank Maggie Bolton and Cathrine Degnen for their comments, and to Piers Locke for carefully reading this chapter. I am, of course, responsible for any errors or infelicities.

Notes

[1] The concept of excorporated blood among the Navajo has been studied by Maureen Trudelle Schwarz (2009), in a discussion of how Navajo consider a donor's blood to continue to live in the body of a recipient following a transfusion. If either partner in the transaction dies, Navajo say that the survivor will suffer at least a partial death (Schwarz 2009: 158). Like Aymara speakers, Navajo prefer to donate blood within the kin group.

[2] The occurrence of domesticated alpacas has been recognised in the archaeological record of the Puna de Junín in Peru from between at least 6000 to 5500 years before present (Wheeler 1984: 402). In an area closer to the herding communities of the south-central Andes studied in this chapter, the domestication of llamas has been recorded for the Atacama Desert as taking place in the fifth millennium before present (Hesse 1982: 11). The meaning of domestication has been explored by Dransart (2002).

[3] The quotations are taken from book 3, chapter 3, paragraph 10 of *An essay concerning human understanding*. The emphases are in the original.

[4] In Aymara, "[u]ka ñuñuyipï, uka ithapu" and, in Spanish, "[e]sa pues le hace mamar, esa placenta pues" (Arnold and Yapita with Tito 1999: 164).

[5] I would also add that the substances under discussion here have complex chains of associations, a case in point being provided by Brandes (1981: 224) who discussed the use of the Spanish term *leche* (milk) in reference to semen.

[6] The original entry is: "No ay nombre generico que los comprehenda todos: a los de quatro pies llaman. Pusicayuni" (my translation).

[7] The original entry is: "tahuachaquiyoccuna se toma por todos los animals, o los que no son hombres" (my translation).

[8] In the English language, these outermost threads are termed the "selvedge" or "selvage".
[9] Bastien (1985: 596) discusses the apparent coincidence of Greek humoral pathology and Qollahuaya concepts of physiology and Ortega Perrier (2001) medieval Christian and Isluga concepts of the soul in relation to European visual imagery. Neither author assumes that European concepts are prototypical of contemporary Andean systems of belief and knowledge; they address the issues on the basis of their fieldwork findings. Bastien (1985: 596) leaves open the question whether Qollahuaya physiology might be pre-Hispanic in origin, derived from Greek humoral theory or arose from a combination of both.
[10] Wind instruments feature in a study of the mutual interdependence between musical and agricultural production in the Quechua-speaking community of Kalankira, Bolivia, in which Stobart (2006: 290) glosses the term *animu* (a Spanish loan word) as "the animating aspect or energy (sound, light, movement) of all living beings or other phenomena with autoresonance, internal movement or which are sources of light".
[11] In this volume, Marcia Stephenson considers European interests in South American camelids.
[12] The information in this section is based on conversations I have had with smallholder owners of camelids in West Wales, many of whose comments echo statements made in specialist magazines such as *The Camelids Chronicle* and *Alpaca: The British Alpaca Society Magazine*. It is supplemented by a brief survey of relevant web sites, both British and North American.
[13] This situation does not occur in the Andes, where the fox is one of the main predators of newly born camelids (Dransart 2002: 28-9).
[14] Sometimes llamas and alpacas are put in a halter for showing them to tourists in urban settings. According to Maggie Bolton (personal communication) an NGO she observed in Lípez, Bolivia, has started to encourage people to adopt both the use of halters and the practice of showing llamas at organised events.
[15] In 2004 the British Camelids Association reported findings obtained from a camelid census of 459 returns. 38% of the owners reported having fewer than two camelids and more than half of the respondents had fewer than four camelids. Owners with more than twenty camelids accounted for 12% of the returns (Brown 2004: 3). Herd sizes in the Andes are variable. Based on a study in Paratía, Department of Puno, Peru, Flores Ochoa (1977) divided alpaca herds into three size groups. "Rich" herders had more than 500 head of alpaca, including herd sizes of 1,000 or 2,000; in the middle range herders had from 300 to 400 head and poor herders had 1 to 20 head of alpaca. The most unfortunate herders had none (Flores Ochoa 1977: 136). In the more arid Puna salada regions of Antofagasta, Chile, Castro Lucic (2000: 100) reported two cases of herd sizes of 200 head in her study, with other people owning herds of the nature of 70, 40 or 10 camelids.
[16] The leaders of a herd are usually mature females without a young animal accompanying them.
[17] Hillel Schwarz (1996: 153) discussed a suite of concepts, including parallel and parallax, containing *para*, "by the side of", or "to one side of, amiss", or "side by side, simulated".

Bibliography

Apaza Suca, N., K. Komarek, D. Llanque Chana and V. Ochoa Villanueva. 1984 *Diccionario Aymara-Castellano Arunakan liwru Aymara-Kastillanu*. Lima and Puno: Proyecto Experimental de Educación Bilugüe –Puno.

Arnold, D. Y. and J. de D. Yapita, with M. Tito 1999 *Vocabulario aymara del parto y de la vida reproductiva de la mujer*. La Paz: ILCA and FHI.

Ayers, M. R. 1981 "Locke Versus Aristotle on Natural Kinds". *The Journal of Philosophy* 78(5): 247-272.

Bastien, J. W. 1985 "Qollahuaya-Andean Body Concepts: A Topographical-Hydraulic Model of Physiology". *American Anthropologist* 87(3): 595-611.

Bennett, M. M. 2002 "Husbandry: the Importance of Headcollar Fit. Solving Major Behavioural Problems … in 30 seconds". *Alpaca: The British Alpaca Society Magazine* 14 (Spring): 18-21.

Bertonio, L. 1984 [1612] *Vocabulario de la lengua Aymara*, La Paz: CERES, IFEA and MUSEF.

Brandes, S. 1981. "Like Wounded Stags: Male Sexual Ideology in an Andalusian Town". In S. Ortner and H. Whitehead (eds.), *Sexual Meanings: the Cultural Construction of Gender and Sexuality.* Cambridge: Cambridge University Press: 216-239.

Briggs, L. T. 1979 "A Critical Survey of the Literature on the Aymara Language". *Latin American Research Review* 14(3): 87-105.

Brown, H. 1948 "Jean Denis and Transfusion of Blood, Paris, 1667-1668". *Isis* 39 (1/2): 15-29.

Brown, J. 2004 "Camelid Census Report". *The Camelids Chronicle* 63 (July): 3.

Burt, S. 1991 *Llamas: An Introduction to Care, Training, and Handling*. Loveland, CO: Alpine Publications.

Castro Lucic, M. 2000 "Llameros de puna salada en los Andes del Norte de Chile". In J.A. Flores Ochoa and Y. Kobayashi (eds.), *Pastoreo altoandino: realidad, sacralidad y posibilidades.* Plural Editores and MUSEF: 85-109.

Cereceda, V. 1993 "Cette Étendue entre l'Altiplano et la Mer . Un Myth Chipaya hors Texte". In A. Becquelin and A. Molinié (eds), *Mémoire de la Tradition*. Nanterre: Société d'Ethnologie: 227-284.

Columbus, C. K. 2006 "Punning in Multi-Modal Narratives". In P. Dransart (ed.), *Kay Pacha: Cultivating Land and Water in the Andes.* Oxford: B.A.R: 115-126.

Cumont, F. 1959 [1922] *After Life in Roman Paganism: Lectures Delivered at Yale University on the Silliman Foundation*. New York: Dover Publications, Inc.

Dahl, G. and A. Hjort 1976 *Having Herds: Pastoral Herd Growth and Household Economy*. Stockholm: University of Stockholm, Stockholm Studies in Social Anthropology 2.

Dransart, P. 2002 *Earth, Water, Fleece and Fabric: An Ethnography and Archaeology of Camelid Herding in the Andes*. London and New York: Routledge.

—. 1996 "Las flores de los rebaños en Isluga: la vida cultural de los ganaderos y camélidos en el norte de Chile". *Nuevo Texto Crítico* Year 9, no 18: 29-39.

Encyclopedia Britannica 1977 *The New Encyclopaedia Britannica*. Chicago: Britannica Inc.

Farr, A.D. 1980 "The First Human Blood Transfusion". *Medical History* 24: 143-162.

Flores Ochoa, J. A. 1977 "Pastoreo, tejido e intercambio". In J. A. Flores Ochoa (ed.), *Pastores de puna: uywamichiq punarunakuna*. Lima: Instituto de Estudios Peruanos: 133-154.

Franklin, S. 2001 "Sheepwatching". *Anthropology Today* 17(3): 3-9.

Gell, A. 1998 *Art and Agency: an Anthropological Theory*. Oxford: Clarendon.

Gegner, L. E. 2000 "Llama and Alpaca Farming, ATTRA (Appropriate Technology Transfer for Rural Areas)". http://www.attra.org/attra-pub/PDF/llamaalpaca.pdf, accessed 8 February 2007.

Gonçalez Holguín, D. 1952 [1608] *Vocabulario de la Lengua General del todo el Perú llamada Lengua Qquichua o del Inca,* Lima: Universidad Nacional Mayor de San Marcos.

Gramiccia, G. 1988 *The life of Charles Ledger (1818-1905): Alpacas and Quinine*. Houndmills: Macmillan Press.

Hann, C.M. (ed.) 1998 *Property Relations: Renewing the Anthropological Tradition*. Cambridge: Cambridge University Press.

Hesse, B. 1982 "Archaeological Evidence for Camelid Exploitation in the Chilean Andes". *Saugetierkandeliche Mitteilungen* 30(3): 201-211.

Homer 1965 *The Odyssey of Homer*, translated by Richmond Lattimore. New York, Evanston and London: Harper and Row.

Ingold, T. 1994 "From Trust to Domination: An Alternative History of Human-Animal Relations". In A. Manning and J. Serpell (eds.), *Animals and Human Society: Changing Perspectives*: 1-22. London and New York: Routledge.

—. 1980 *Hunters, Pastoralists and Ranchers: Reindeer Economies and their Transformations*. Cambridge: Cambridge University Press.

Locke, J. 1975 *An Essay Concerning Human Understanding*, ed. Peter H. Nidditch. Oxford: Clarendon Press.

McLaughlin, M. M. 1974 "Survivors and Surrogates: Children and Parents from the Ninth to the Thirteenth Centuries". In L. de Mause (ed.), *The History of Childhood*. London: Souvenir Press (Educational and Academic) Ltd: 101-181.

Mallon, J. 1998-2002 (revised 2006) "Gentling & Training Llamas & Alpacas: Aberrant Behaviour Syndrome (ABS)". http://www.mallonmethod.com/llearning4.html, accessed 8 February 2007.

Mamani Mamani, M. 2002 *Diccionario práctico bilingüe Aymara-Castellano. Zona Norte de Chile. Suman chuymamp parlt'asiñi*. Antofagasta: EMELNOR NORprint.

Ortega Perrier, M. 2001 "Escatología andina: metáforas del alma". *Chungara (Arica)* 33(2): 253-258.

Ritvo, H. 1987 *The Animal Estate: the English and Other Creatures in the Victorian Age*. Cambridge, Mass.: Harvard University Press.

Schiebinger, L. 1993 "Why Mammals are called Mammals: Gender Politics in Eighteenth-Century Natural History". *American Historical Review* 98(2): 382-411.

Schwartz, H. 1996 *The Culture of the Copy: Striking Likenesses, Unreasonable Facsimiles*. New York: Zone Books.

Schwarz, M. T. 2009 "Emplacement and Contamination: Mediation of Navajo Identity through Excorporated Blood". *Body & Society* 15(2): 145-168.

Stamberg, G. and D. Wilson 2007 "Berserk Male Syndrome". Llamapaedia http://www.llamapaedia.com/problems/bms.html. Accessed 8 February 2007.

Stobart, H. 2006 *Music and the Poetics of Production in the Bolivian Andes*. Aldershot, Hants: Ashgate, SOAS Musicology Series.

Strathern, M. 1988 *The Gender of the Gift: Problems with Women and Problems with Society in Melanesia*. Berkeley and London: University of California Press.

—. 1984 "Subject or Object? Women and the Circulation of Valuables in Highland New Guinea". In R. Hirschon (ed.), *Women and Property, Women as Property*. London: Croom Helm: 158-175.

Szemiński, J. 1987 *Un kuraca, un dios y una historia*. San Salvador de Juyjuy: Sección Antropología Social (ICA), Facultad de Filosofía y Letras, Universidad de Buenos Aires.

Tani, Y. 1996 "Domestic Animal as Serf: Ideologies of Nature in the Mediterranean and Middle East". In R. Ellen and K. Fukui (eds.) *Redefining Nature: Ecology, Culture and Domestication*. Oxford and Washington, D.C.: Berg: 387-415.

Torres Rubio, D. de 1616 *Arte de la lengua aymara*. Lima: Francisco del Canto.

Wheeler, J. C. 1984 "On the Origin and Early Development of Camelid Pastoralism in the Andes". In J. Clutton-Brock and C. Grigson (eds.), *Animals and Archaeology 3, Early herders and their flocks*. Oxford: BAR International Series 202: 395-410.

Willis, R. 1990 "Introduction". In R. Willis (ed.) *Signifying Animals: Human Meaning in the Natural World*. London: Routledge: 1-24.

Wood, C. T. 1981 "The Doctors' Dilemma: Sin, Salvation, and the Menstrual Cycle in Medieval Thought". *Speculum* 56(4), 711-727.

Yolton, J. W. 2001 "Locke's Man". *Journal of the History of Ideas* 62(4): 665-683.

CHAPTER FOUR

ANIMAL KNOWLEDGE: THINKING THROUGH DEER AND SHEEP IN KYRGYZSTAN

STEPHANIE BUNN

What I am concerned to address in this chapter is how Kyrgyz animal knowledge can contribute to our understanding of the relationship between animals and science, or even possibly offer an alternative framework for thinking about what can and cannot be discussed under the rubric of science. Furthermore, I ask, if this is the case, then how far can we take this line of enquiry? To do this, I am going to focus my discussion specifically on Kyrgyz food, and examine the way that social sciences consider that animals are "good to think", specifically with reference to Kyrgyz relations with deer and sheep.

The Relationship between Traditional and Empirical Knowledge

The positivist scientific approach in general, most especially in the hard sciences, is usually concerned with using experimental methods to test theories and hypotheses through repeatable experiments. The setting up of experiments as practised in the hard sciences, however, is not a methodology that I have encountered Kyrgyz pastoralists using to inform their understanding of animals. The naturalist method in the sciences, including some aspects of biology, on the other hand, examines organisms in their "natural" state and habitat, relying upon observed, empirical, practical research to produce rigorous and reliable results. It is this empirical approach, a kind of qualitative research through lived-experience among animals, which forms the basis of my view that it is possible to incorporate Kyrgyz knowledge of the animals they work with into Euro-American scientific understanding of these animals. This is, after all, a methodology which has also formed the basis of our approach to the ethnographic research of human societies. My aim in this chapter, therefore, is firstly to examine the kinds of empirical knowledge that Central Asian pastoralists in general, and Kyrgyz pastoralists in particular, have developed about animals. I will then explore how this knowledge, gained through social practices and bodily experience, relates to other kinds of knowledge that Kyrgyz people have developed about animals, such as knowledge that anthropologists might call "symbolic". My aim is to explore what this can tell us about the boundaries which anthropologists perceive to exist between the scientific and the symbolic.

Kyrgyz Pastoralism and Science

The animal knowledge of Kyrgyz pastoralists is based on repeated, observed and lived experience, developed and communicated over 2,500 years of Central Asian nomadic pastoral practice. Archaeological evidence (Rudenko 1970, Gryaznov 1969) tells us that from the age of the Early Nomads (500BC) onwards, nomadic pastoralists in Central Asia have employed very similar forms of food production and have husbanded animals and their environment in broadly similar ways. They have bred similar, although developing, forms of animals for the same sorts of animal products and for textile production. Early Nomads also engaged with wild as well as domestic animals, as do today's Kyrgyz and other Central Asian pastoralists. Alongside archaeological material, animal terminology concerning the five main herded animal names is evident in expressions and metaphors in the Kyrgyz and Kazakh languages, reflecting a long standing, well-honed, relationship with animals incorporated into all aspects of everyday life, including speech (Mawkanuli 2002).

This chapter focuses on food, deer and sheep, so I am going into less detail on people's working relationships with animals than I would like. Nevertheless, a very brief summary is necessary (this can be read in greater depth elsewhere [Bunn 2000, 2003]). Through fieldwork and material gathered from informants, my view is that Kyrgyz pastoralists' relations with animals involve a greater emphasis on mutual cooperation, and are often described in terms more akin to social relations than is the case with our European practices of husbanding animals. This can be seen in diverse practices from bringing home sheep, which is less directive than European methods, often delegated to goat leaders, and yet achieves full acquiescence from the herd; milking, where dependent offspring are kept attached to the mother, so the effect is one of cooperation; eagle training, where the trainer conceives of the bird in terms of partnership; and in killing domestic or wild animals, when people pray to the souls of despatched animals to ensure the prosperity of their families. In this, I am at variance with Ingold, whose work suggests that pastoralist relations with animals are of control or domination (Ingold 1994, Bunn 2003). In my view, among Central Asian pastoralists, this is certainly not the case, and animals, domestic or wild, are accorded the same attitude in treatment (see also Tani 1996).

In its engagement with societies which have prioritised a scientific and rational approach to animal husbandry over the past 100 years, Kyrgyz practical, empirically tested local knowledge has been rarely regarded as contributing to scientific knowledge. During the Soviet period, for example, Soviet scientific advances were used to justify breeding programmes for the production of cross-bred sheep with higher meat value, to encourage twin lambing, and to introduce a variety of new veterinary medical practices. While benefits undoubtedly accrued from the use of new medicines for sick animals in regard to illnesses such as foot and mouth disease, for example, and in avoiding illnesses such as brucellosis being passed from animals to humans, this "scientifically-based" input was not always beneficial. The new crossbreeds, such as local breeds with Merino sheep, were often unsuited to both the high mountain pastoral environment and to local needs for non-food animal products such as wool. Increased flock sizes due to twin lambing also led to overgrazing and degradation of high mountain pastures. As Farkas (2002) argues, local knowledge in Inner Asia had its own way of managing certain epizootic diseases, while some other diseases, such as brucellosis have only been an issue since cows were introduced to the region following Russian annexation in the 1870s.

If Soviet approaches to "animals and science" in Central Asia tended to undervalue local empirical knowledge, more recent approaches by foreign NGOs have not always fared much better. Indeed, many of the issues raised by anthropologists of environmentalism in other regions can be seen in the recent global interventions with Kyrgyz environmental and animal management. National parks and conservation areas are being proposed which do not always incorporate peoples as living parts of the local ecosystem, while new animal produce, such as full-fat milk cheeses not always to local taste, are being developed for export by Swiss companies, and so on.

The Soviet approach to Central Asian pastoralists' knowledge was undoubtedly based on a view developed through the Russian Academies of Sciences that nomadic pastoralists were "pre-logical", and not possessing a rational, scientific mode of thought. This is encapsulated in the work of the Soviet psychologist Luria (whose research among Kyrgyz and Uzbeks in the 1930s [1976] forms the basis for Walter Ong's (1982) later theories on orality and literacy). Luria went so far as to say that because Kyrgyz and Uzbeks did not have writing, they lacked the ability to reason, the prerequisite, in his view, for scientific thought (1976). Luria carried out extensive research among Uzbeks and Kyrgyz in Central Asia, testing them with syllogisms and other exercises. When his respondents refused to draw general conclusions from assertions made by the researchers without checking the facts for themselves, he condemned them as being incapable of rational thought. Today we would hopefully view such experimental approaches to human cognition as productive of their own answers, and the answers received as reflecting the reluctance on the part of the respondent to accept what the interviewer said at face value.

However, if the assumptions made by Soviet scientists about nomadic mental capabilities now seem themselves illogical, it is a great irony that development workers from esteemed international organisations today, at times, make apparently similar implicit assumptions in their earnest desire to develop exports of animal products from the region or to "conserve" the local environment. It is not the intention of this paper to explore the continued introduction of new farming practices and environmental projects in Central Asia further, however. My intention is to consider Kyrgyz animal knowledge in greater depth. In the process, I hope to show that aspects of Kyrgyz animal knowledge, distilled through a long history of repeated practical experience within quite finely focussed parameters, should be given credit for extending our knowledge of animals both in terms of science and possibly beyond.

Food Science

I will begin my exploration of Kyrgyz animal knowledge with an examination of animals as food. The diet of Kyrgyz pastoralists working in the mountain pastures is relatively restricted in comparison with Euro-American expectations. Animal products from sheep, horse and cow provide the main sources of nutrition, usually in combination with homemade pasta and bought-in rice or potatoes. These are supplemented by occasional fresh vegetables brought up by visiting relatives, wild onions and other gathered food, and some hunted meat.

However, what we might see as an apparent limitation in diet, restricted to mainly animal products, results in a progressive and extensive focussing of awareness on the potential and properties of those animal products, resulting in an in-depth understanding of how to use them, how they affect the human body, and a unique approach to what food does. In what follows, I argue that the experience of eating and ingesting animals as food provides Kyrgyz pastoralists with an important practical and empirical knowledge about animals through the embodied experience of incorporating animals into themselves as food.

i) Milk Products

The variety of products Kyrgyz women make from cows' and horses' milk is legion. Cows are milked twice a day. Almost all this milk is processed. Each time, the cows' milk is first scalded (to protect against brucellosis and other diseases) and immediately processed into full-fat milk (*ak süt*), skimmed milk (*kök süt*), and cream (*kaymak*). People rarely drink fresh cows' milk, except in tea in cold weather in the mountains, and even then, they often prefer boiled butter.

Secondary products include yoghourt (*ayran*) from white milk; cream cheese (*süzmö*) from blue milk (that is, fully skimmed milk), which may be made into hard cheese balls which keep over winter (*kurut*); clotted cream skimmed from the scalded milk (*körmöchök*); boiled butter (*sarymai*); and a kind of fudge made from the residue of boiled butter (*chöbögö*). Shakhanova, describing Kazakh milk products, cites the high nutritional value of these processed foods, *kurut* for example containing 53% protein, and 3700 calories per kilo, –very important for a winter diet (1989: 112). The value of sour milk products such as yoghourt includes their live bacterial culture/probiotic content which has recently become fashionable in Euro-American "healthy" diets.

Kyrgyz use of cows and their milk products is comparatively recent, since cows were mainly introduced by the Russians over the past two centuries, the most common type being the Latvian Brown (Emeljananko 1994). The increased use of cows' milk parallels the decreased use of camels' and yaks' milk over a similar period, as these animals became replaced by cows, and were used less frequently as pack animals with the increased introduction of motorised transport. *Shubat* (fermented Bactrian camels' milk) and dried cheese balls made from yaks' milk were also formerly popular milk products, both of which have a very high calorific value.

Mares' milk does continue to be used, and the best-known product made from it is *kumiss*, (fermented mares' milk). However, there is also the first milking of the mare each day, which is drunk with relish for its health-giving properties and contains a concentration of hormones. This may possibly be related to the use of pregnant mares' urine in HRT.

Kumiss was used by the "Early Nomads" from the 5th Century BCE (Shakanova citing Potapov, 1989, 113). Shakhanova notes that 1 pint of *kumiss* contains almost all the requisite vitamins for human needs in a 24 hour period, and that it possesses antibiotic properties which counter E. Coli (*Ibid:* 113). Levine further advises that mare's milk is particularly valuable for human diet because of its essential fatty acid composition (1998). A combination of *kumiss* with boiled mutton broth (*bouillon*) is used as a very effective post-parandial digestive–an excellent stomach settler following a meal of *besh barmak*, the Kyrgyz celebratory meal of a whole boiled sheep.

So, what may appear to the outsider to be a simple diet of milk products and meat is for Kyrgyz pastoralists a complex diet, varied and finely tuned, that reflects an in-depth knowledge of the food potential of animal products. In terms of milk products, alongside its food potential, there is also a clearly held empirical understanding of how these foods can affect the well-being of human body, as illustrated through the use of the first milking of mares and the use of a mix of fermented mares' milk and broth as a digestive. Thus food is not just mundane bodily fuel, its capacity to affect bodily wellbeing in positive or negative ways is always considered alongside its nutritional capacity and certain kinds of food are considered good for the health under certain circumstances and may be used medicinally.

ii) Meat and Fat

This knowledge of food properties and its use in regard to both nourishment and wellbeing is also applied to meat and fat. Meat is categorised as either hot or cold. People I met said that lamb and mutton meat and fat were cooling, while horse meat and fat were heating foods. Butter and oil are also cooling. The Soviet ethnographer, Shakhanova cites evidence to show that horse meat contains significantly more carbohydrate than other meat, and that horse fat is more easily absorbed than other animal fats (1989: 112). For this reason, horse meat is especially valued as food in winter, and a sausage made from the fat is particularly popular, is very warming, and is also given to people recovering from illnesses who are feeling weak. (Some was sent to me from Kyrgyzstan by a Kyrgyz friend when I was recovering from major surgery, and it was certainly most nourishing). Furthermore, because the horse is a simple stomached, non-ruminant grass eater, its meat and fat contain essential linoleic fatty acids, which Levine advises are very important for foetal growth (1998).

Kazakhs use this kind of fat for weaning babies. Shakhanova also advises that these acids are essential to prevent skin disease (1989: 112).

Mutton is eaten all year round and is generally more digestible (Emeljanenko 1994). This, Levine argues, is owing to the fact that sheep (and cows) are ruminants (1998). They have complex stomachs, resulting in their fat containing a higher proportion of mono-unsaturated fats and their meat is therefore more of a standard. The cooling properties of sheep's fat are also considered to assist in physical illness, and the fat is rubbed on to the chest in case of fevers. The power of fat extends to its use in cooking, and the aroma of bread (*borsoc*) fried in sheep's fat, is ascribed an especially enriching power which is one reason why it is cooked on special and celebratory occasions.

There are several possible explanations of this approach to categorising food as "hot" and "cold". Ibn Sina, who was one of the founding fathers of modern medicine, lived in west Asia in the tenth century and also analysed food in terms of the production of heat and coolness on the body (Halm 1997, 69). Migration and conquest among nomadic people in this region have resulted in many cross-cultural influences. In the Mongol era (thirteenth to fourteenth centuries), Muslim, Chinese and traditional Central Asian foods were all eaten at the court of Kubilai Khan (Buell 2000). The contacts between peoples of this era were such that the Chinese produced the book "*Muslim medicinal recipes*" based on Ibn Sina's *Qanum fi al-tibbi* (Buell 2000).

From my own experience and from what people have said to me in the field, I would also suggest that such terminology reflects a practiced local knowledge of the potential and properties of animal products for bodily well-being, and is a result of an embodied understanding through accumulated experience of ingesting animals in this way. Ethologists now argue that certain animals seek out the most appropriate plants to eat if they are sick. It would be surprising if human societies had not developed the same kind of knowledge about their food in a more refined form. It is also entirely possible, given that local oral histories convey information from many centuries ago as if it were quite fresh, that Ibn Sina's ideas had either been influenced by, or had influenced pastoralists' animal knowledge.

From a structural anthropologist's perspective, the physical properties of food might be considered irrelevant, and the categories "hot" and "cold" might be interpreted as classic binary oppositions, and used to help decode specific social practices. What I am arguing, however, is that firstly, Kyrgyz people's experience of food is gained through lived experience and through ingesting food, and that secondly, their knowledge of food extends beyond that of simple food-as-bodily-fuel, to one which perceives food to supply both physical wellbeing and health alongside bodily nourishment.

The Concept of *Küt*

Thus far, what I have described regarding Kyrgyz knowledge of food might be accepted as constituting a kind of empirically gained "folk science" attained through lived experience and repeated practice. However, a difficulty arises when we try to impose the kinds of defined limits that Euro-American scholars normally perceive exist between what we would call a "real" or empirical understanding of food and what is often described as a "symbolic", ritual or metaphorical aspects of food and human-animal relationships. Such boundaries do not always exist at the same points in Kyrgyz categorisation.

One Kyrgyz word for food is *küt*. (This is quite an old-fashioned word, although still spoken in Kyrgyz. It is the opposite of *jüt* which means famine, or a time of pestilence [Farkas 2002, Bunn 2008]). Along with "food", *küt* also means "happiness", "nourishment", "prosperity", "the power of life" or "soul/spirit", and is also one of several names for the "Pole Star" (Yudakin 1965, Shakhanova 1989). Among Kyrgyz and Kazakhs, human experiences such as happiness and prosperity are considered to be able to be incorporated into material objects, in people and in food. For example, if a woman is leaving her family for a short while, there would be concern that she will take away her personal stock of happiness when she leaves. This problem can be averted by transferring the woman's happiness to the food prepared to be served at a celebratory meal before she goes. By eating the meal served, the woman's family will ingest her stock of happiness–"the agency of the food enables the happiness to be returned to its source" (Shakhanova 1989: 116). When I first did fieldwork, one informant said I should have brought some of my mother's food with me so that I would not miss her. Ironically, I had brought with me some biscuits which she had made, which I had been saving for a lonely day. This provided a clear confirmation, in my Kyrgyz friend's view, of my agreement with them on this subject. In contrast, when a Kyrgyz person is near death, the family will remove all food from the tent or room where the dying person lies, in case it absorbs the person's spirit when they die.

Although this very tangible understanding of quite abstract qualities of food could be interpreted as a sensory evocation of memory, it is also applied to colour and at times equates sensory experience with social relations. Whiteness, for example, indicates a sacred (*y'yk*) aspect in many spheres of life and, with regard to food, the whiteness of milk products is on a par with the whiteness of bone or shell, giving them a peculiarly sacred aspect among Central Asian and Mongolian pastoralists. *Kumiss* is scattered to the four directions when setting up a new tent for example, or as an offering to powers in nature in the event of disaster (Humphrey with Onon 1996: 79). Other food products also have quite literal attributes and are used accordingly. Sweet food, for example, whether sugar cubes, sweets, or milk products and bread, will be eaten at weddings, funerals and at matrimonial negotiations to help sweeten relationships and proceedings.

The importance of this extended range of values that are attributed to food is such that at *toy*, or celebratory feasts for lifecycle events, food, especially animal products, take on a significant ritual role. I have already mentioned that one reason for cooking bread (*borsoc*) in hot mutton fat at feasts, is that the smell of fat is considered to promote well-being. It is also why *ploff*, or *pilau*, a mutton and rice dish cooked in fat (the only dish to be prepared by men), is so popular at weddings. This beneficial smell of fat does not just extend to the living, however. At "memorial feasts" for the dead, women strive to cook food in hot fat, especially mutton fat, in order that the ancestors (*arbak*) may also feed upon the aroma and thus bring their good will to bear upon the proceedings. And indeed, cooking food with fat at any feast will encourage the ancestors' auspicious presence. One friend said that thinking of his deceased relatives inspired him with a sense of "debt" which could only be assuaged by cooking bread in hot mutton fat. Privratsky discusses similar practices among Kazakhs (2001). Likewise, when moving into a new house, people will rub mutton fat along the doorframe to ensure the wellbeing of the future inhabitants. The smell of fat ensures wellbeing not just of the living inhabitants, but through attracting the ancestors, can bring both their presence and their goodwill.

Most anthropological thinking on this subject would prefer to categorise this second aspect of Kyrgyz knowledge about animals and food under the heading of "belief". That would usually mean, symbolic, but not *true* or *scientifically verifiable*. But this categorisation is ours, not theirs. The science: not science, i.e. "belief", opposition with which we would tend to classify these different aspects of nourishment, would not necessarily be considered salient by Kyrgyz herders, since there is no necessary locally defined difference between people's knowledge of the nourishing effect of meat and fat on the body, its ability to promote wellbeing and happiness, or the capacity of its aroma to call down the ancestors, for example.

The Kyrgyz concept *küt* is linked to a wider integrated paradigm, denominated by Yukata Tani (1996) as the Central Asian "ideology of nature". Humphrey refers to a similar way of thinking about the world among Mongolian herders (1996) and it is described by Mongolian anthropologist Tseren (1996) as "balance with nature". My own field informants called it "harmony with nature". This is a paradigm developed about the wider environment, not just through thinking about the world, categorising and conceptualising it, but through lived experience of it. Descriptions such as "harmony" or "balance" may sound romantic (although as Andrew Whitehouse in this volume comments, such terms can equally often appear in the vocabularies of scientists). I would argue that they reflect a notion of nature as more of a self-organising system, with different players of which humans are one participating element. In this case, other players might include 'animal masters' or 'powers in nature' such as those that animate sacred sites or *mazars*. Such an appreciation of nature may be expressed both in the general working relationship that people have with animals and in terms of animals as food. This kind of approach to nature, as Humphrey (1996), Hamayon (1990) and also Vitebsky (2005) describe for the wider Inner Asian and Siberian region, is part of a way of living with the environment associated with the past practice of shamanism.

According to Hamayon, who follows the approach of Russian anthropologist Basilov, the practice of hunting is directly related to human health and well-being, because hunting animals involves the repayment of human souls for animal souls (1990). The shaman's role is to negotiate with animal masters or mistresses for as many animals as possible in the hunt, but this abundance must repaid in some form, if not by illness and death within one's own community, then from another. Hence the importance of balance, the relationship between hunting and health, and quite possibly the sense of a need to repay debt to the ancestors through feeding them with cooked hot mutton fat. Although there are few if any practicing shamans as of old these days among either Kyrgyz or Kazakhs, many attitudes and understandings about the integrated relationships between powers in nature, animals and the environment, which were a part of this former shamanistic understanding of the world, are still held, and this includes understandings of food. As Vitebsky says (2005), every time northern hunting people throw vodka on the fire to the spirits of the hearth they are "doing shamanism", and as Humphrey describes, every time Mongolian herders scatter milk to the four directions or to the earth and sky (1996), or, I would suggest, when Central Asian people feed ancestors with the aroma of fat at feasts, such understandings are enacted.

Human, Animal and Social Relationships

Thus far I have discussed Kyrgyz animal knowledge with special reference to food, exploring how what seems to be an experiential, practical knowledge of what food does to the body through eating, at times crosses the boundaries that we may perceive to exist between science and not-science. In this section, I would like to move into other areas of social experience, specifically the relationships which Kyrgyz people consider to exist between themselves and sheep and deer. The relevance of these relationships for a discussion of "animals and science" is that it is illustrative of the wider paradigm in which Kyrgyz animal knowledge is situated.

i) Sheep Bones and *Besh Barmak*

One Kyrgyz practice which reflects the specific philosophy of nature outlined in the previous section is the apportioning of a killed sheep at the main Kyrgyz celebratory meal known as *besh barmak.* Among Kyrgyz and other Central Asian pastoral peoples, a "cut" or killed sheep for a ceremonial meal is divided into very specific portions which are known in Kyrgyz as *ustukans*. These *ustukans* are associated with specific family and status relationships, the expression of which is re-enacted every time a sheep is killed and served at feasts. When a sheep is cut on such occasions, a prayer is said to Kuday (God) to ask for a blessing, prosperity and plenty for the family who provided it. The meat in effect becomes sacred, or *y'yk*.

The way the dead sheep is cut up into specific *ustukans* is very reminiscent of the way that hunting peoples cut up their kill. The joints are precisely divided, according to family and local practice, and each one has a name. The whole sheep is prepared for the meal, *ustukans*, intestines, head, legs, and so on, and is boiled for about three hours. In fact, one of the traditional measures of time is the time it takes to prepare such a meal.

Guests at such a meal are first treated to the "tablecloth"–*dastarkhan*–given tea and breads fried in sheep fat (*borsoc*), as described above. The most significant moment however begins when the first tablecloth is removed at the start of the second course and the boiled sheep cut into *ustukans* is brought in. The guests in the tent, or room in the house, are already seated in precisely determined positions according to age, sex and status, and according to the points of the compass if in a tent, which is usually erected facing south. Men are on the left of the tent, looking in (west) and women on the right (east). Oldest and most prestigious persons sit in the *tör*, or place of honour, at the back of the tent, youngest such as the daughter-in-law, the *kelin,* and young horsemen, *jigids*, sit near the door. Objects are positioned according to a similar system. Rooms in houses are set out in an equivalent arrangement for feasts, although orientation is less relevant. It is essential that everyone is seated correctly in relation to everyone else. To be seated incorrectly is something people remember for life.

When the *ustukans* are brought into the tent, a young boy will first bring round a jug of water for washing hands. He is wished a *bata*, blessing, by each individual in turn, and the *ustukans* are then distributed according to the same system of seniority and relatedness as people are seated in the tent. The two buttocks, *jang bash*, have highest status, and can be given to men or women of highest social standing. These are followed by the *kuyrik*, the pelvic area, including the vagina, of a female sheep, which always goes to the oldest and most prestigious woman. (It can never go to a man.) Then comes the upper thigh or *joto jilik*, then lower thigh or *kashka jilik*, middle bones or *orto ustukan*, ribs or *kabygan*, the forearm or *karjylyk*, the shoulder blades or *daly*, the *künjilik* which is a special *ustukan* reserved for the daughter-in-law, or *kelin,* who will be serving the meal, and the breast or *tüsh* which is reserved for young girls and women, daughters of the host. Like being seated incorrectly, to be given an inappropriate *ustukan* is something one never forgets. Along with the *ustukans*, the sheep's head is brought in, and it too is cut up (by a respected young man) and is apportioned according to his perception and his hopes for relations with others. The eye may be given to someone he wants to look into more deeply, the palate may be given to someone he considers creative, or artistic, and so on.

One might argue that this way of relating the relative status of people to the body of the sheep also raises the possibility of thinking through animals in a structuralist way, which raises a kind of "social science" opportunity for the anthropologist. Correspondences with animal body parts extend also into the wider environment as their names are also given to features of the landscape. In terms of thinking through animals, this is as Caroline Humphrey says, "elementary structures country" (1995: 144-5). Furthermore, the very word for bone, *sö'ök*, is also the word for relative or relation, and across Central Asia, among both the Kazakhs and Mongols, (although not among Kyrgyz) people are divided into the "black bone" (common people) and "white bone" (aristocracy).

Bones also have a whole range of uses beyond the way they are distributed at feasts linked to family wellbeing, prosperity and protection, especially for male family members. Sheep's shoulder blades are used in divination, sheep's knucklebones (*chükö*) are given to young boys for luck, the elbow to shoulder bone (*karjylyk*) is kept and used to protect sons, ram's horns are hung over the doors of tents and houses for prosperity, and so on. Bone terminology also extends beyond human-animal relationships and is also used for defining features of the landscape. There is also a clear use of bone imagery in traditional Kyrgyz felt pattern design (Bunn 2005, 2008).

In relation to how bones are distributed at feasts, I would argue, however, that this is not simply a case of using sheep's bones, and indeed the concept of "bone" in general, as a means of classifying kin, or a kind of "thinking through sheep" metaphor. I suggest that the use of "bone" in creating and recreating social relationships at meals is part of the ongoing complex of relationships between humans, ancestors, animals, and the wider environment, which is manifest in the Kyrgyz philosophy of nature, outlined above, which considers the destinies of people, animals and also features of the land as intertwined. In terms of "animals and science", the range of relationships with animals, from working with them, incorporating them into the body, to the social and metaphysical, cannot be separated into clear categories of "real" versus "symbolic" or "metaphor".

When Kyrgyz *ustukans* are distributed at a formal meal, they are rarely given to the same person on each occasion, but are apportioned according to who is there at the time, and the position in the tent of each person relative to others. That is, *ustukans* are not given to specific persons, but are distributed according to the relationships of people within the gathering. This process is relational not fixed. What people are doing when they share and eat the cooked *ustukans* between themselves is articulating their relationships between one another in relation to the group present, as well as in relation to the wider environment in terms of the orientation of the tent and to the body of the sheep. They are distributing and ingesting the sacred (*y'yk*) quality of the meat of the animal (which has been blessed when killed) according to how humans are related to each other, to how a sheep's body is considered to be divided or ordered, and to the orientation of their immediate tent or room environment. In other words, what we have here is a living, changing matrix of perceived ongoing relations between humans, animals and the wider environment, not a set of fixed metaphors or symbolic categories. The whole process relates to the dynamism of social relationships, not to fixity, and it is as much about incorporation of the sacred meat as it is about the unfolding relationship between humans, animals and environment. Perhaps this fluidity is also reflected in the way that Kyrgyz nomenclature for kin is much more flexible than our own kin terms (which are very fixed). Kyrgyz kin terms are "relative", for example, biological uncles might be called brothers, or cousins, or father according to their comparative age in relation to an individual.

The extension of this set of unfolding relationships between people and animals also incorporates the environment and can be illustrated through the appreciation of sacred places, or *mazars,* and the powers and terminology they share between them. A sacred place might be a feature of the landscape such as a spring, grove of trees, waterfall, *or* the tomb of an ancestor, a relative, or Sufi saint. The spirit of the place, *e'e*, might be conceived of as an animal, a deer or a white camel (Bayaleva 1972), a natural spirit *or* a person. Spirits of place can, for example, also be called *ata*, or father, and this term is particularly used to refer to mountains. When Kyrgyz people meet for the first time, one of the frequent opening gambits in conversation is to ask of each other the names of their seven patrilineal ancestors, a relationship known as *jeti ata* (seven fathers). This practice of referring to one's seven patrilineal ancestors, is mirrored in the landscape where seven key Kyrgyz mountains are also named *jeti ata*, (names vary from place to place). The Seven Stars, the Plough constellation, are likewise Seven Old Men, protective ancestors in the sky (Heissig 1980), (but the Plough is also called Jeti Arkar (seven wild sheep) (Yudakin 1965), all indicative of the interweaving of relationships between people, ancestors, animals and the wider environment.

The use of the term bone, *sö'ök*, for both bone and relative, the interrelated roles that ancestors and features of the environment can have at sacred sites, the different beings who can take on the role "father" (ancestors, mountains, sheep or stars), and the way that kin terms can apply to different relations in a much more flexible and "relative" way to our own are all resonant with notions raised in Strathern's discussion of *The Relation* (1995). Strathern's concern is how generative terms such as "relation" can have different, yet associated, abstract and concrete meanings. In my understanding, Strathern puts forward the case for a kind of "intellectual synaesthesia" or resonance, a merging of terminology, for apparently different aspects of human behaviour which in fact converge, such as the concrete "kin relation" and the more abstract "social relationship". I'd suggest that in the same way as we rarely describe the full sensory range of particular unified, synaesthetic experiences of the world, but may use terms from one sense (such as sight) to refer to a complex sensory experience, so when Kyrgyz herders describe their relations with each other through sheep "bone", these are not necessarily mutually exclusive kinds of relations, but are part of a complex, multi-faceted unfolding set of interconnected relationships, where human destinies are perceived of as linked to those of ancestors, animals and the wider environment, all of whom are active players

ii) Bugu *Enyé*

Of the Northern Kyrgyz, two of the most noted *chong uru'u* ("tribes") are Bugu (Reindeer) tribe and Sary Bagysh (Yellow Elk). There are many stories, especially among Bugu, tracing their ancestry to a deer woman with horns, known as Bugu *enyé*, "Deer Mother". Around Karakul at the eastern end of Lake Issyk-kul, the motherland of Bugu, there are kiosks decorated with images of Bugu *enyé*, small *chai kanas* (tea houses) called after her, and so on. Abramzon (1978) and Bayaleva (1972) recorded people saying, "*Bizdin enyebiz tubu bugunun kyz ekim.*" (Our foremother is the daughter of the deer), and many stories are told linking her to Bugu people.

One story goes that two hunters saw a beautiful woman in the midst of a herd of deer. They carried her off and married her to their brother's son, Alseit, a famous Kyrgyz ancestor. Before she would marry him, she laid down three conditions–that he must never call her "hunting trophy bride"; that he must never hit her on the head; and that he must never look at her naked body.

The woman had two horns which she renewed each year. That's why they called her Muyuzdu Baybiche, "the woman with horns", or Bugu *enyé*, "the Mother of the Deer". One day, Alseit broke the ban on looking at her body, and he saw her completely transparent, with all her internal organs completely visible. She died at that moment and later disappeared. (Bayaleva 1972).

On first examination, these references to deer ancestors, Bugu (Deer) tribes, bodily prohibitions and fantastic images of skeletal animals could be interpreted as a reflecting an implicit totemic system relating kin to deer. A more extensive examination however, reveals that the whole phenomenon of the deer ancestor is far wider than just among Kyrgyz, extending across Northern Asia to groups such as the Evenk (see Ansimov 1963), where the term *Bugadu enentin* refers to the clan mother, the mother of hunted animals and protector of the hearth in the image of the female deer. This is the mistress of all hunted animals in North Asia, the final recourse to which "a shaman will go in appealing for help to find a catch" (1963).

The skeletal body of the deer woman is also evocative of the practice among Siberian and Central Asian hunting peoples, including Kyrgyz when they hunt deer and wild Argali sheep, of cutting up the dead animal in a prescribed way and allocating trophies to participants in the hunt, akin also to the practice of "breaking a stag" in European and American hunting, and in much the same way as Kyrgyz domestic sheep are cut into *ustukans*. Kyrgyz herders are also hunters, and cloven-hoofed wild animals such as deer and wild sheep are sometimes called *Kudaydyn maly*–God's herds, suggestive of a special relationship with hunted animals.

I'd argue that the relationship between people and deer is no more "totemic" than that between people and sheep, but that both examples illustrate the textures and relationships which adhere within the Kyrgyz philosophy of nature, outlined above, where the destinies of people, animals and place are inextricably interwoven. Thus while a structuralist or a symbolist might argue that the deer too is good to think, I would suggest that in Kyrgyz relations with animals, deer are not symbols or metaphors, but key players in a world view where there is a perceived interconnectedness between people, animals and the environment. Sheep and deer are relations–sheep through the patriline, while deer are "mothers".

Animals and Science and Animal Knowledge

The theme of the conference which inspired this chapter was "Animals and Science". The intention behind the conference, as I understand it, was to see how a focus on animals could contribute to anthropological studies of science, or knowledge in general. I wonder if this theme does not also draw our attention to a century old concern with the "nature-culture" debate, especially totemism, and the more recent concerns with "human-animal relationships" (Ingold 1994, Tani 1996). While I do not want to go into a deeper debate about whether or not there is such a thing as totemism, certain aspects of the totemism debate could help illuminate the question of how to deal with the wider framework of Kyrgyz animal knowledge.

Lévi-Strauss, whose work on totemism rather drew a line under the subject over 40 years ago, viewed it as "an artificial unity, existing solely in the mind of the anthropologist, to which nothing specifically responds in reality" (1962: 79). Yet in the first chapter of *Totemism* he also takes great pains to define totemic phenomena, and of course, later in the book, to perform a structuralist analysis of them. He generally agrees that totemism is usually taken to comprise a set of *relations* between two series, one cultural, the other natural, for example, a kin or social group (culture) and a species of animal or plant (nature), so that the perceived connexions between humans and animals are used to qualify sets of relatives (1962: 79). This may or may not involve food prohibitions or attributions of sacredness to that species. Having made this explicit, his concern is not to explain these relations in Euro-American scientific terms, but to explore their diversity, how their differences illuminate certain aspects of social relations and to decipher or decode them (Poole 1962: 29).

However, in a sense, Lévi-Strauss's refusal to deal with the "why" of totemism, while liberating him to make structural analyses of the relationships within the practice, has at the same time resulted in a denial of certain aspects of the relationships between the humans and animals involved. For him, the relations themselves are "symbolic", yet what they "represent" or "are" is not relevant in comparison with how they fit into a wider set of structural relations. This is compounded by the necessity of structuralism to require that the players or manifestations of totemism are significant only as part of a "code", but are not meaningful in themselves, but only in relation to other aspects of the phenomena. Furthermore, his dismissal of the work on the subject of anthropologists such as Durkheim(1915), Malinowski (1948), Fortes (1945) and others, because of their concern with *explanation* through affectivity, utility or symbol, has somewhat consigned their ideas to obscurity, when one could argue that their work too has value, if not for enhancing an appreciation of totemism, then at least for possibly elucidating the affective aspects of human-animal relations.

In this attempt to see what we can learn from Kyrgyz animal knowledge, I would suggest that, probably since the inception of "totemism" in anthropology, human-animal relations have been addressed on the one hand, as practical, as in people's working knowledge of animals through husbanding, herding, hunting, or eating, and on the other, as "symbolic" or "metaphorical", that is as vehicles which carry meaning, for example through what was called totemism. In both the structuralist approach and that of more symbolist approaches, the dualism created in thinking this way has meant it has been impossible to understand the relationships or relations involved as the unity perceived by the people concerned (Descola and Pálsson 1996).

Most importantly, I would argue, each aspect of the relationship between humans and animals must be perceived for what it is, a relationship mediated through the intersubjective understandings between humans and animals through working with them, communicating with them and the sensory experience of eating and ingesting them. Working and communicative relationships with animals include the processes of herding and being herded, training and being trained, milking and being milked, riding and being ridden, hunting and being hunted, and killing and being killed, and involve sensory, intersubjective, and even social exchanges. Food and the eating of animal products include the experience of ingestion, taste and smell, digestion and nourishment. These relationships and experiences are not just "mundane", physical or pragmatic, but have quality and can be extended into the social, the ancestral or the sacred. Therefore the social or sacred understandings people have of animals are not purely metaphoric or symbolic, but involve bodily experience and resonance. In my interpretation of Kyrgyz animal knowledge, these aspects of the relations between animals and people are therefore, contra Levi-Strauss, meaningful in themselves, and can be affective. This is not to deny the value of a structuralist interpretation of this material, but rather to state its limits. It is therefore also important to explore how the experiences of such relationships work and what limits of knowledge are involved.

Conclusion: Relations and Relationships

How does this help our discussion of animals and science? Firstly I argue, obviously, for the inclusion of local knowledge within scientific analyses of animals. Central Asian pastoral practice includes a long developed, finely-honed, empirical knowledge of animals and their products, reflecting an extensive historical interdependence between humans and animals in the region. This includes how to manage them in the mountain and steppe environment, how to process them as food, and also what they do to the human body. It also reveals the existence of a slightly different "unit of analysis" in regard to food, in comparison with the Euro-American pragmatic "food as fuel" approach, since across Central Asian pastoral groups, their knowledge of food does not separate its use for nourishment, physical well-being and even medicine. Secondly, I argue that Kyrgyz animal knowledge does not necessarily employ the same dualistic science/non science divide that our science does between itself and other forms of knowledge, and that stretching the boundaries may be useful for furthering knowledge of the "science" of animals. Seeing others through our own dualistic categories of science: not science and husbandry: totemism may limit our ability to appreciate the world views of others as integrated systems, and prevent us from engaging with their "intellectual landscape" (Descola and Pálsson 1996), something which were we to do could conceivably broaden our "scientific" understanding of the relationships between humans and animals in general. Most specifically, when talking of human-animal relations, the relational aspect must be considered. Kyrgyz pastoralists' "relations with animals" are based on strong working relationships with them which are affective, although not romantically so, but which certainly lead to a kind of understanding, and even an emotional bond, between both parties. This leads me to my third point which is that when considering human-animal relations, by including people's experience of animals as food, as social and as sacred, an expanded and potentially very productive understanding of peoples' knowledge practices could be developed which would enhance what we might gain from a more classical perspective,

Kyrgyz animal knowledge reflects the very integrated understanding of the environment which Kyrgyz and other Central Asian and Mongolian pastoralists have, which incorporates science/non-science; people/animals/ ancestors/environment; mundane/sacred into one coherent system. The understanding of food as *küt* and illustrations of Bugu *enyé*, and eating *besh barmak* reveal, not simply examples of symbol systems, but a clearly perceived and experienced connectedness between humans, animals and environment. Can we address such a unified way of being in the world through the lens of science? Not to do so would seem to limit our understanding of these specific ways of knowing the world. An engagement along the boundaries of our different paradigms could be productive.

Bibliography

Abramzon, S.M. 1978 "Predmety Kulta Kazakhov, Kirgizov i Karakalpakov (The Subject of the Cults of the Kazakhs, Kirgiz and Karakalpaks)". In *Materialnaya Kultura i Xozyastvo Narodov Kavkaza, Sredney Azii i Kazakhstana: Sbornik Museya Antropologii i Etnografii No. XXXIV* (Material Culture and Economics of the People of the Caucasus, Central Asia and Kazakhstan: Collection of the Museum of Anthropology and Ethnography No. XXXIV). Leningrad: Nauka.

Ansimov, A.F. 1963 "Cosmological Concepts of the Peoples of the North". In H Michael (ed.), *Siberian Shamanism*. Toronto: University of Toronto Press: 157-229.

Bayaleva, T.A. 1972. *Doislamskiye Verovaniya i ikh Perejitki u Kirgizo* (Pre-Islamic Belief and its Survival among the Kyrgyz). Bishkek/Frunze:Ilim.

Buell, P. 2000 *A Soup for the Qan: Chinese Dietary Medicine of the Mongol Era as Seen in Hu Szu-Hui's Yin-Shan Cheng-Yao.* London and New York: Kegan Paul International.

Bunn, S.J. 2008 *Nomadic Felts*. London: British Museum Press.

—. 2005 “From the Domestic to the Divine: Kyrgyz felt textiles in central Asia”. *Toronto Studies in Central and Inner Asia* No 6.

—. 2003 “The Relationship Between Land and People in Rural Kyrgyzstan”. *Toronto Studies in Central and Inner Asia* No 6.

—. 2000 *The House of Meaning: Tents and Tent Dwelling among the Kyrgyz of Kyrgyzstan*. PhD thesis: University of Manchester.

Descola, P and G. Pálsson. 1996 “Introduction”. In P. Descola and G. Pálsson (eds.), *Nature and Society*. London and New York: Routledge: 1-22. Durkheim, E. 1915 *The Elementary Forms of the Religious Life*. London.

Emeljanenko, T. 1994 “Nomadic Year Cycles and Cultural Life of Central Asian Livestock: Breeders Before the Twentieth Century”. In C Leeuwen, T Emeljanenko and L Popova (eds.) *Nomads in Central Asia*. . Amsterdam: Royal Tropical Institute: 37-68.

Farkas, O. 2002 “Reinventing the *dzud*: livestock famine in 21st century Mongolia”. *Toronto Studies in Central and Inner Asia.* No 5.

Fortes, M. 1945 *The Dynamics of Clanship among the Tallensi.* Oxford: Oxford University Press.

Gryaznov, M.P. 1969 *South Siberia*. London: Cresset Press.

Halm, H. 1997 “The Land beyond the River: Islam in Central Asia”. In J Kalter and M Pavaloi (eds.), *Uzbekistan*. London: Thames and Hudson.

Hamayon, R. 1990 *La Chasse a l'Ame : Esquisse d'une Theorie du Chamanisme Siberien* Nanterre: Publications de la Societé d’Ethnologie.

Heissig, W. 1980 *The Religions of Mongolia*. Los Angeles: University of California Press.

Humphrey, C. 1995 “Chiefly and Shamanist Landscapes”. In E. Hirsch and M. O'Hanlon (eds.), *The Anthropology of Landscape*. Oxford: Clarendon Press: 135-162.

Humphrey, C. with Onon, U. 1996 *Shamans and Elders.* Oxford: Clarendon Press.

Ingold, T. 1994 “From Trust to Domination”. In A. Manning and K. Serpell (eds.), *Animals and Human Society: Changing Perspectives*. London: Routledge: 1-22.

Levi-Strauss, C. 1962 *Totemism*. London: Beacon Press.

Levine, M. 1998 “Eating Horses: the Evolutionary Significance of Hippophagy”. *Antiquity*. 72 (275): 90-111.

Luria, AR. 1976 *Cognitive Development: Its Cultural and Social Foundations*. M.Cole (ed.) Trans. M Lopez-Morillas and L Solotaroff. Cambridge: Harvard University Press.

Malinowski, B. 1948 *Magic, Science and Religion.* Boston: Doubleday Anchor

Mawkanuli, T. 2002 "Nomadic Culture and the Kazakh Language". *Toronto Studies in Central and Inner Asia.* No 5.

Ong, W. 1982 *Orality and Literacy: the Technologizing of the Word.* London: Methuen.

Poole, R. 1962 "Introduction" In C. Levi Strauss, *Totemism.* London: Beacon Press.

Privratsky, B. 2001 *Muslim Turkistan.* London: Curzon.

Rudenko, S. 1970 *Frozen Tombs of Siberia.* London: J.M. Dent and Sons. (First published 1953 as *Kultura Naseleniya Gornogo Altaya v Skifskoe Vremya.*) Moscow/Leningrad: Academy of Sciences of the USSR.

Shakhanova, N. Zh. 1989 "The System of Nourishment among the Eurasian Nomads: the Kazakh Example". In G. Seaman (ed.), *Ecology and Empire,*. Los Angeles: Ethnographics Press.

Strathern, M. 1995 *The Relation.* Cambridge: Prickly Pear Press.

Tani, Y. 1996 "Domestic Animal or Serf: Ideologies of Nature in the Mediterranean and the Middle East". In R Ellen and F Fukui (eds.), *Redefining Nature.* Oxford: Berg: 387-415.

Tseren, P.B. 1996 "Traditional Pastoral Practice of the Oirat Mongols and their Relationship with the Environment". In C Humphrey and D Sneath (eds.), *Culture and Environment in Inner Asia.Vol 2,*. Cambridge: The White Horse Press: pp147-159.

Vitebsky, P. 2005 *The Reindeer People* London: Harper Perennial

Yudakin, K.K. 1965 *Kirgiski-Russkii Slovar.* Moscow: Sovietskaya Entsiklopedia

CHAPTER FIVE

THE BALANCE OF NATURE: ENTANGLED SCIENCE AND ETHICS IN DEBATES ABOUT RAPTORS IN SCOTLAND

ANDREW WHITEHOUSE

Balance and Imbalance

During my doctoral fieldwork on the Scottish island of Islay, I often spoke with people about the animals they noticed around where they lived. One group of animals that were mentioned more than most were the various birds of prey, or raptors, that could be seen in Islay. Sometimes my informants spoke negatively about these birds, whether they were buzzards, hen harriers, peregrines or golden eagles. Sometimes they spoke with great awe and admiration. But, in whatever manner they were discussed, it seemed that these birds were regarded as important and powerful, in terms of both their physical presence and their symbolic freight. Looking more broadly at discourses of nature conservation and the countryside in Britain, raptors again figure prominently and, in part, the aim of this chapter is to explain why these birds have become such a pivotal concern to so many groups involved in the management and use of the British countryside. These debates are exemplified by the subject of this chapter – the scientific study of the population ecology of raptors and grouse conducted at Langholm Moor in the Scottish borders during the 1990s.

Two of the most prominent groups involved in debating raptors and grouse in Scotland have been nature conservation organisations and sport shooting organisations. For both, the identification of ecological and economic balance is integral to their aims and it is these claims that form the subject matter of this chapter. "Balance" is employed by these parties to represent a positive outcome of human actions. Appropriate actions are envisaged as creating an ecological balance that maintains stable populations of important but competing species. In doing this, competing economic and political demands are understood to be reconciled and the appropriateness of particular patterns of human-environment relations is reinforced. What constitutes appropriate human-environment relations reveals important moral differences between conservationist and sporting approaches, differences that have a long history in interactions between the two. This chapter explores these differences and examines the role of science in attempts to reach reconciliation and "balance".

As outlined above, the idea of balance plays an important role in the discourses of both conservation and sport shooting organisations. More recently, balance has also emerged as a key term within anthropology, particularly as applied to relations between humans and their environment. Writers focussing on this subject have, for example, adapted Sahlins' trope of balanced reciprocity to understand the contrast between paternalistic and orientalist ecological relations (Pálsson 1996) or to identify "environmentalist cultures" (Milton 1996). In a famous case, Rappaport (1967, 1971) argued that relations between humans and their environment can be self-balancing and that the mechanisms for maintaining ecological balance are often cultural. Using the example of cycles of pig feasts in Papua New Guinea he attempted to show how the timing of feasts was critically related to ecological and social pressures. These cycles thus operated as cybernetic systems that incorporated self-balancing mechanisms analogous to the governor of an engine (see also Bateson 1972, 1979). In all of these approaches balance is understood to be desirable in human-environment relations and actions, values and beliefs are considered to have an important influence in regulating ecological relations. Anthropological approaches to human-environment relations have thus tended to theorise ecological balance as an outcome of the same sorts of moral behaviour that, in an earlier era, were considered to maintain social cohesion.

Herein I explore the rhetorical employment of balance and the role that science is taken to play in revealing both balance and imbalance. Balance is a great example of a "hurray term" and, as such, it has great potential for employment in moral and political claims about how the world ought to be. Balance is employed to convey a desirable state and the invocation of imbalance is to point to a state of affairs inherently in need of rectification. When claims of balance are made, it means that circumstances are right, appropriate and normal and implies the reconciliation of differences and oppositions. When balance is looked for in nature, science is seen to play a critical role in its assessment, a role that is assumed to have consequences for political and moral action.

Raptors and Gamebirds

Prior to a discussion of the implications of the Langholm research, a brief resumé of raptor conservation in Britain over the past hundred years or so is necessary. In short, most species did badly until around the 1970s but have done much better since, to the extent that species such as buzzard, peregrine and sparrowhawk, which had all become rare or localised during the twentieth century, have more recently become widespread and sometimes common (Holloway 1996; Snow & Perrins 1998). The explanations given for the scarcity of raptors for much of the twentieth century are twofold. Firstly, gamekeepers killed them, either directly or indirectly, for reasons that will be discussed later. Secondly, because raptors were at the top of the food chain, many were very badly affected by organochlorine pesticides such as DDT: scientists demonstrated that these chemicals reduced the breeding success by causing their eggshells to thin, consequently making the eggs more vulnerable to damage (Prestt 1966; Ratcliffe 1970)[1]. Laws were subsequently introduced banning the use of organochlorine pesticides and outlawing the direct and indirect killing of raptors. Since these laws were introduced, raptors have indeed done much better. The increases and expansions in raptor populations over the past few decades represent a significant triumph for British nature conservation organisations, particularly for the Royal Society for the Protection of Birds (RSPB), a large charitable conservation body that has built much of its reputation on campaigning for the protection of birds of prey. It also represents a triumph for the use of science in conservation because science established the link between organochlorines and raptor populations and scientists have also successfully argued that a healthy raptor population is indicative of a healthy ecosystem rather than an imbalance. Science and conservation have “stood in” for raptors and won the argument on their behalf (see Campbell this volume; Yearley 1993). Science was used to reveal the effects of human activity on a particular aspect of nature and, once these effects were established, political conservation was able to mitigate the deleterious effects of these activities through regulations. In this case it was established in law and policy that raptor populations should neither be directly controlled by humans nor indirectly controlled by the use by humans of chemicals, and these actions were prohibited and policed. Legitimacy was given to a particular model of human intervention in nature, in which manipulation of animal species was shunned in favour of manipulation at the level of habitat.[2] Raptors were re-imagined as indexical of good quality habitat and appropriate human-environment relations, rather than being seen as indicative of neglect or ecological imbalance.

This last view has long been prominent within the world of field sports where raptors are widely, although not universally, believed to have a deleterious effect on the economic viability of game shooting. The aim of management for sport shooting is to ensure the maximum number of gamebirds, such as grouse, for the autumn shooting season. An important aspect of achieving this has been to control the numbers of predators so that man, in effect, becomes the principal predator of the birds. This was the reasoning behind the widespread killing of raptors on sporting estates that led to the rarity of many species during the twentieth century. Although few within field sports would now publicly advocate the indiscriminate and illegal killing of birds of prey that had once been widespread,[3] there have been calls for the legalisation of the licensed control of raptor populations. In Islay many of my informants, including both gamekeepers and farmers, believed that raptors had become too numerous and that this was causing declines in many other bird species. These included some of the species that conservation organisations such as the RSPB were working to protect on the island, such as the corncrake and the chough. Conservationists meanwhile have remained vehemently opposed to any softening of the more-or-less blanket protection that birds of prey receive in law.

The debate that has emerged between the industries of conservation and sport shooting can thus be seen as representing different views on the degree and manner in which humans intervene in nature. Nevertheless, for both sides what remains crucial is the maintenance of what they perceive as an ecological balance. My intention in this chapter is to examine further these different identifications of balance as they relate to birds of prey and to consider their reflections in the different positions adopted by conservationists and the shooting fraternity in the debates developing from the Langholm Project.

The Langholm Study

Farmers I spoke to in Islay often mentioned birds of prey in connection with their effects on other species. When I asked a farmer why he thought some wildlife had declined, he replied:

> These days there's no control over birds of prey because there are no gamekeepers and the place is inundated with ravens and buzzards. So now there are fewer small birds and fewer of the little ducks around the loch. There's an imbalance, with RSPB demanding nature be left to take its

> course and sort things out. But nature needs a hand to keep things in check. If things were left to do as they please then the place would be over-run by red deer. I had a friend who was very knowledgeable about all the wildlife and he argued all his life against the RSPB and tried to point this imbalance out to them, but they wouldn't listen.

So this farmer was putting forward an apparently straightforward explanation about the relationship between humans, predators and prey. The gamekeepers had once controlled the numbers of raptors and crows but the decline of sport shooting in Islay meant that now there were fewer keepers. At the same time the RSPB had successfully demanded that birds of prey be fully protected. The result of these two changes meant that raptors increased dramatically and that other species suffered because, quite obviously, they were being preyed upon by the raptors. This perhaps seems unarguable but, as the farmer's deceased friend had discovered, the RSPB hold a rather different view of the relationship between predators and prey. They have argued for the apparently paradoxical position that birds of prey are not harmful to populations of their prey species. They have even argued that this is the case when the raptors are generalists, birds that feed on other species and so can survive the extinction of one species of prey, and when the population of the prey species is vulnerable.

This was a position the RSPB adopted in relation to the study of raptors and their prey conducted at Langholm. The Langholm study was part of a larger project called the Joint Raptor Study (JRS) and was conducted between 1992 and 1997 (Thirgood *et al.* 2000: 99-100). The research was headed by two biologists, Simon Thirgood and Stephen Redpath,[4] and was funded by a consortium consisting of both conservation organisations, including the RSPB and Scottish Natural Heritage,[5] and shooting bodies, including the Game Conservancy Trust and sporting estates. A number of areas were studied as part of the project but the most significant site was Langholm Moor, which had long been an important area for red grouse and for grouse shooting. The landowner, the Earl of Dalkeith, agreed that no illegal killing of raptors would take place at Langholm, although the numbers of other predators, such as crows, would continue to be legally controlled. The experiment was thus premised on the estate working within the law and only controlling predators that were not protected by legislation. According to the Scottish Gamekeepers' Association (SGA) the Langholm project:

> Proposed to resolve the age-old argument about whether birds of prey and the grouse they hunted could survive together if there was no interference from human beings… The argument went back and forward, but it was a

> sterile one, because both sides had only prejudice to go on (Scottish Gamekeepers Association 2001: 9).

So here, in quite bald terms, is one particular take on the debate and the role that science was intended to play in resolving it. In the first place, the debate concerned the relations between raptors and grouse in a "state of nature", where "there was no interference from human beings".[6] It should be emphasised that "nature" within this debate is not synonymous with the idea of "pristine wilderness", as it often is in conservation issues outside of Britain. The British landscape, to both conservationists and the sporting industry, has attained both its diversity and biodiversity through long-term human activity, which means that human intervention in nature is a given, at least within Britain. As such, this debate typifies the modern British anxiety about the degree and type of intervention in nature (see Campbell this volume; Strathern 1992). Secondly, the debate took place in the human world in which "only prejudice" was possible without the use of scientific study. Science has the special capacity to study the world of nature and then report to the human world about the facts that are uncovered there. These assertions follow the problematic relations between science on the one hand and politics and ethics on the other outlined by Bruno Latour in *Politics of Nature* (2004), which builds on ideas introduced in his *We Have Never Been Modern* (1993).

Latour draws on the Platonic narrative of "the cave", in which humanity metaphorically resides, unable to see the world of reality, with its non-human laws, clearly. The task of the philosopher, and later the scientist, is to journey from the social, human world of the cave to the reality outside. This fable thus creates a contrast between the social world and science as forms of knowledge and order. It is only science that can gain knowledge of the reality of nature. But it is also asserted that politics only operates within the human world of the cave.

The narrative was consolidated through the seventeenth century debate between Hobbes and Boyle about the remits of science and politics, the former concerned with matters of fact within the controlled environment of the laboratory, the latter concerned with wholly human affairs from which non-humans are excluded (Latour 1993). So, in the Old Regime, or the "Modern Constitution", politics and nature have been kept separate through works of "purification", with only science able to move between the two. Latour calls this "the Two House Collective" or the "Bicameral Collective" (2004: 37). Latour's beef is not with science itself but with how it is represented by "the epistemology police" and he thus marks a distinction between what he calls "Science" and "the sciences" (*Ibid*: 10-18). To Latour, Science is a kind of ideology, linked to the outmoded ideas of the cave, and serves to destabilise politics through its presentation of natural facts, whereas the sciences are the separate disciplines concerned with taking account of non-humans and enabling their incorporation into the collective.

At Langholm, the experiment of non-intervention at the level of animal species was established and populations of grouse and raptors, most notably hen harriers and peregrines, monitored in order to understand how numbers of grouse and raptors were interrelated if no human interference was involved. Science, in the SGA's view, was expected to provide clear facts about nature that would override the prejudicial values of human interest groups. As the experiment unfolded, the changes in the populations of all these species came as a surprise to many. Hen harriers increased quite markedly, from two breeding females in 1992 to twenty in 1997 and peregrines increased from three to six pairs (Thirgood *et al.* 2000: 99). During the same period the number of grouse shot on the moor declined to such an extent that in 1999, two years after the study ended, grouse shooting was abandoned by the estate, which then had to lay off its staff of gamekeepers. Harrier numbers then dropped sharply during the late 1990s and only two females remained by 2003 (The Game Conservancy Trust 2003). During 1995 and 1996 the numbers of grouse taken by harriers and peregrines were also monitored (Thirgood *et al.* 2000: 100). These figures suggested that, had raptors been absent from the moor, there would have been four times the number of grouse at the start of the shooting season in mid-August.

These figures in themselves suggest a significant link between raptor populations and grouse populations. An increase in raptors is followed by a decline in grouse and a subsequent decline in raptors. But biologists argued that there was more to these circumstances than a simple causal relationship between increased raptor numbers and declining grouse stocks. In particular, they invoked two other factors both of which were related to the quality of habitat at Langholm (Thirgood *et al.* 2000). Studies of long-term numbers of grouse at Langholm indicated that the annual bag had declined throughout the twentieth century. Over most of this period raptor numbers were very low on the moor because of the attentions of gamekeepers so it seemed that raptors were not to blame for the long-term decline in grouse numbers, even if they had had some effect during the 1990s when they were more numerous. Instead, it was the scientists' view that the change during the twentieth century that most influenced grouse populations was a decline in the extent of heather moorland and a consequent increase in the acreage of grassland. This was caused by a sinister enemy of the grouse: the sheep.

Sheep tend to graze out heather and this has the knock-on effect of increasing the populations of voles and meadow pipits, both of which do well in grassland areas. These two animals are the main prey species of hen harriers, which also feed on grouse. So an increase in sheep numbers reduced the area of heather cover at the expense of grass. This meant that there was less suitable habitat for grouse but more for other species preyed upon by hen harriers. Once illegal killing was removed, the harriers did well because there was so much food for them. The grouse, which had long been declining, become so scarce that shooting was no longer economic. Thus the conclusion of the biologists was that the most important factor influencing the population dynamics of raptors and grouse was habitat and this was a position strongly supported by the RSPB, who in an article from their members' magazine asserted that sheep were "the real enemy" (Royal Society for the Protection of Birds 1998). Conservationists and scientists went on to suggest that a monoculture of heather would be better for grouse than a mixture of grass and heather and that this would limit harrier numbers so that a better balance between the two would be achieved (Thirgood *et al.* 2002). The right sort of intervention in habitat could create an ecological balance between predators and prey that would mean human intervention, in terms of controlling raptor numbers, would be unnecessary in order to maintain grouse shooting. However, research also showed that when densities of grouse are low, either because of habitat changes or because of the dynamics of grouse populations, raptor predation *can* limit grouse populations and thus reduce the economic viability of shooting (Thirgood *et al.* 2000: 100). Whilst it emerged that habitat changes were behind long-term declines in grouse, changes in habitat take a long time to achieve and, in the meantime, it was recognised that harriers and peregrines were seriously reducing the economic viability of shooting at Langholm, even if they were not threatening the survival of red grouse on the moor in the longer-term. In other words, both conservationists and the sporting lobby found support for their views on appropriate forms of human-environment relations in the findings of the scientists and so each were able to make political claims that they had won the argument even though compromises might be made on policy.

Various proposals were put forward as to how raptor numbers could be balanced with the economic requirements of grouse shooting. These included the translocation of raptors from grouse moors to moors where shooting did not take place (Thirgood *et al.* 2000: 102; Watson & Thirgood 2001) and the diversionary feeding of raptors so that they avoided preying on grouse (Amar *et al.* 2004; Redpath *et al.* 2001;

Thirgood *et al.* 2000), a solution that it was hoped could represent a compromise between the different interests (Redpath *et al.* 2004). Nevertheless, the method of intervention that some shooting groups advocated was the control of raptor numbers. Some gamekeepers remained unconvinced by the arguments against this approach put forward by conservationists and some scientists in response to the JRS. The SGA argued that:

> By 1997, there was little doubt about the outcome… What had once been one of the most successful moors in Britain had ceased to be viable as a commercial proposition. Given the terms of the experiment, the RSPB should, at this point, have accepted the conclusions and got together to find means of reducing the harrier numbers… Instead of this, however, the RSPB simply moved the goalposts. The decline of grouse, they said, had nothing to do with raptors. It was due to the poor state of the moor, the lack of heather cover, and the way the Earl's sheep had grazed the hills over the years… Not surprisingly, the owner protested. None of this had been raised at the outset, and grouse had always flourished alongside the sheep in this supposedly poor habitat. Who was to pay for the moor now? With no grouse to shoot, and therefore no income, he would have to lay off his keepers. Surely, the point of the experiment had been to find a solution, not to destroy the only local source of employment (Scottish Gamekeepers Association 2001: 10).

So for the SGA the conclusions to be drawn from the Langholm data were clear. Raptors had clearly been the main reason behind the cessation of grouse shooting on the moor and their numbers needed to be controlled. Furthermore, this was actually better for the raptors because, without shooting, the moor would no longer be managed properly and so the quality of the habitat for raptors would deteriorate. But, the SGA argued, because the RSPB seemed unwilling to admit to being wrong, they tried to argue that the supposed deterioration in habitat quality was to blame. Indeed the scientists' claims that the habitat had worsened were questioned by a former keeper at Langholm, who argued that management of the heather had always been appropriate and intensive. Without this management, and the grouse shooting that pays for it, the countryside of the hills and moors would be devastated. As the SGA put it:

> The scenery so cherished by tourists and ramblers alike–rolling purple hills, rocks and golden bracken–is in fact preserved by gamekeepers, who know how to maintain the balance of wildlife in the countryside and who are able to sustain biodiversity and employment in rural areas by creating a shootable surplus and farming that crop carefully. Take away one element–

> the grouse–and all that is at risk. Try arguing that with the RSPB however and they will talk about poor habitat and overgrazing. What you will not hear is anything about the human beings who have lost their jobs, the small birds and grouse that have gone, and the countryside that has been denuded–all in the name of so called "conservation" (Scottish Gamekeepers Association 2001: 11).

Grouse are thus revealed as essential to maintaining a balance between different elements within nature and to reconciling oppositional requirements of biodiversity conservation and economics. Firstly, an ecological balance needed to be achieved between predatory raptors and their prey of grouse. Secondly, and indexical of the first, a balance was required between the economic concerns of the grouse shooters and the desire of conservationists to protect important species. One might gloss this, as the environmental historian Chris Smout (2000) has, as a balance between use and delight, or as Campbell (this volume) argues, between a modernist productionist logic and a late-modern emphasis on lifestyle and leisure. Gamekeeping is here proposed as the means through which humans intervene in such a way as to regulate all relationships, both socio-economic and natural, to appropriate levels. Grouse are thus seen as indicators of the reconciliation of ecology and economy and of appropriate human-environment relations. The RSPB disrupt this system, it is claimed, because they do not attend to the grouse as an economic resource that, in the SGA view, achieves balance both between and within society and nature. Instead they look to raptors as indexical of habitat quality and of a balance in the populations of predators and prey that emerges as a result, provided human interference in those populations is constrained

Entangled Science and Ethics

Within the foregoing, science and ethics appear to be entangled in the debates generated by the Joint Raptor Study. Science has revealed evidence concerning the interactions of grouse and raptors and this in turn has illuminated the effects that human interest groups have on these birds and their habitats. As a consequence, it also foregrounds ethical questions of how the land should be and the ways in which humans should intervene in nature. Just as the new technologies of "pharming" and genetic modification (see Degnen and Campbell this volume) have provoked debates about what is valued in our relations with animals, so has the new evidence provided by scientific research into raptors. It was hoped by the SGA that Science would resolve ethical differences between different positions by presenting ecological facts about nature that would make one of those positions redundant and the other unquestionably correct. This reconciliation of difference was never achieved, however, and controversy burgeoned ever more strongly. Both parties utilised the findings of the scientists to fulfil their own political ends and bolster their ethical positions on human-environment relations.

The SGA argued that science was required in order for each side in the argument to have more than "prejudice to go on". A Latourian "Science" was supposed to make one side right and the other wrong in their debate about the appropriate form of human intervention in nature. But, whatever conclusion was reached, the answer would be rendered in terms of scientific proof rather than ethics. For both sides, there were to be no prejudicial ethics here but simply a system of legislation and best practice based upon knowledge rather than values. A thus rationalised economic and political policy then acts as a proxy for the ethics of human–animal relationships and Science is implicated within this concealment. There was no point in arguing that one configuration of appropriate human-environment relations was more ethical than another unless these prejudices could be given the appearance of fact by Science. But, although there was a desire to distinguish facts and values, what transpired was their entanglement and this in turn resulted in competing claims on policy makers.

Conservationists claimed that an ecological balance should be achieved through habitat management. A central tenet of conservation is that the right habitat leads to the right species at the appropriate densities. Get the habitat right and an ecological balance will surely follow in the long-term. In the case of Langholm, humans needed to intervene in nature at the level of vegetation to create the appropriate conditions for both grouse and raptors to thrive in. Nature, with the right kind of intervention, was thus viewed as inherently balanced and much is made of the "best practice"–the management required to maintain a natural balance. Whilst this tendency towards ecological balance is understood to be immanent in nature and not created by humans, it can be upset by inappropriate human activity. This notion of balance in nature as separate from but indexical of human behaviour is critical. According to conservationist arguments, the spanner-in-the-works of ecological balance was the economic interest of the grouse shooters. Raptors could never extirpate the grouse but it had to be acknowledged that they could have an economic impact on grouse shooting when grouse densities were at a low level because of habitat changes. Conservationists thus came to see human economic interests as out of balance with behaviour that was ecologically appropriate. This required intervention of a different kind to habitat management, for example the practice of diversionary feeding, but the conservationists could never countenance the killing of the birds they had striven hard to protect and that represented many of their claims to knowledge, political influence and legitimacy.

For the grouse shooting fraternity, gamekeeping is the "best practice" for maintaining a balance between "use and delight". Keepering involves intervention at the level of the animals themselves and not just the habitats they live in. The SGA put their case for the solution to what they called "the raptor problem" thus:

> The solution to our problem as we see it are changes in the law to allow for the control of raptors to maintain an ecological balance. Some form of quota system or relocation program needs to be seriously investigated if the raptor problem is to be solved to everyone's satisfaction (Scottish Gamekeepers Association n.d.).

So, for the SGA, the law was inadequate because it prohibited the appropriate degree of intervention required to create a balance between predator and prey and to reconcile competing human interests.

In wanting to affect policy, the conservationists and the grouse shooters had something in common. In fact, the interpretations that both made of the Langholm study were always geared towards policy change. The RSPB argued that the appropriate habitat management required the necessary economic systems to support it.

> The main problems have to do with money and complexities of land use policy. Without better financial incentives for moorland management, including shepherding and keepering, and a change in the system of subsidies so that sporting use competes on a level playing field with livestock farming and forestry, the future for some grouse moors and the birds and people who depend on them may be bleak, with or without raptors (Royal Society for the Protection of Birds 1998).

For the RSPB, then, the scientific conclusions were not in doubt and the problem and its solution lay in the economic and political system enabling the appropriate degree of intervention in nature. Although they have a different interpretation of the research to the SGA, what they do with it is rather similar: they take it to government where they hope to wrestle for favourable changes in economic policy and legislation.

So the science of the Langholm study was open to interpretation and, like a lot of science, raised as many questions as it addressed. It failed to resolve a clear distinction between the facts of nature and the prejudicial values of society, which some at least hoped it would clarify. Instead, the science and ethics of human-environment relations remained entangled.

But, following Latour (2004: 91-127), this entanglement should be seen as integral to the work of the sciences and its articulation with politics. Indeed such hybrid agglomerations, he argues, are an inevitable product of the modern tendency to purification, that is to distinguish nature and society, science and politics (1993: 1-12). Greater clarity emerges when considering the Langholm study in terms of two different powers, both containing notions previously understood in terms of facts and values.

Latour (2004:103) argues that two qualities normally attributed to facts are that they create new uncertainties to the established order but that, conversely, they also end discussion as they become "matters of fact" or institutions. Facts promote discussion but also end it. Values also possess two somewhat opposed qualities (*Ibid*: 105-108). First, they put forward those that *ought* to be consulted, both humans and non-humans. Values are thus concerned with consultation. Second, values are concerned with the compatibility of new propositions with the existing order. Values are placed hierarchically within collective life. Latour goes on to note that facts and values share two opposing powers. First, there is the power to take into account, containing the requirements of perplexity and consultation. Second, is the power to put into order, containing the requirements of hierarchisation and institution. So these new powers capture elements previously contained in facts and values. Their entanglement now comes to seem more purposive and inevitable, even if this is not how the debate is seen by its protagonists.

The Langholm study of course does not exist in isolation from a long-term process of perplexity, consultation, hierarchisation and institution but it adds further perplexing facts into this process that have led to the complications and controversies that Latour sees as integral to the process of bringing nonhumans into collective life. Raptors and grouse and the facts that emerge about their relations have long been subject to processes of consultation and tests to their compatibility with the existing order. The Langholm study and its new facts at first generated confusion and perplexity but this in turn provoked calls or reminders for consultation. Habitat changes and sheep needed to be taken into account as did the numbers of grouse, the beauty of the countryside and the wider ecological conditions of moorland. This led to claims about how the scientists' findings could be integrated with the existing order in which birds of prey were protected and valued and where grouse were integral to the economics of sporting estates. Alternative possibilities were examined and controversy raged on, as it still does. The position of the institution has remained elusive but the calls for policy changes put forward by both conservationists and shooting interests in a sense reflect an acknowledgement that all parties' requirements have a validity that requires some sort of solution to be arrived at through politics. The sciences have shown that habitat change has contributed to declines in grouse but it has also shown that raptors can reduce their populations at low density to levels that are uneconomic. These ecological imbalances that raise questions about human-environment relations have produced new requirements to hierarchically regulate, balance and put into order the interests of shooters, conservationists, grouse, raptors and habitats.

So the question remains of whether this debate represents a "problem", whether ecological, political or social. In a Latourian sense, that there *is* a debate is not in itself a problem, but the terms of the debate seem problematic. The notion of ecological balance is at the heart of these problematic terms but is also symptomatic of wider problems in how relations between humans and animals, science and politics are assumed to operate. Balance is assumed to be at once intrinsic to nature and also indexical of human behaviour. Its presence is both transcendent and dependent. It is nothing to do with humans and everything to do with them. Such an instance of hybridity, a conjoining of conservationists, scientists, grouse shooters, grouse, harriers, sheep, heather and data, is an inevitable product of a Modern Constitution, one that can construct nature and politics in these paradoxical ways (Latour 1993: 29-32). Whilst the aim of resolving a balance in nature encourages attention to our relations with animals and habitats, it is also a kind chimera that can never be clearly resolved or described, not least by the scientists. Like the equally problematic and somewhat related notion of social cohesion, ecological balance can only exist in a kind of laboratory-induced stasis in which both debate and change have ended. Debates would not end if the notion of ecological balance is done away with, but they would be founded on terms that are more visible and describable, something that can aid in taking full account of the multiplicity and complexity of relations, both ecological and political.

Notes

[1] These two scientists went onto become, respectively, Director of the RSPB and Chief Scientist at the Nature Conservancy Council. As such they were amongst the most prominent British conservationists of the 20th century.

[2] It should be noted that British conservation organisations do occasionally still control the numbers of certain animals, when this is legal. This tends to be exceptional rather than integral, as is the case in management for field sports.
[3] A gamekeeper I spoke to in Islay told me he preferred the modern culture of keeping to the indiscriminate "killing for the sake of it" that was the norm when he had started in the job in the 1970s.
[4] These two scientists were employed, respectively, by The Game Conservancy Trust, a charitable research organisation supported largely by people involved in field sports, and the Centre for Ecology and Hydrology, a research organisation that is largely publicly-funded.
[5] The governmental conservation body in Scotland.
[6] Despite such implications that the Langholm study was a closely controlled "experiment" in the classic sense of Boyle, the scientists conducting the research asserted that, "[It] was not a controlled experiment, so it cannot establish cause and effect" Thirgood, S., S. Redpath, I. Newton & P. Hudson. (2000: 100).

Bibliography

Amar, A., B. Arroyo, S. Redpath and S. Thirgood. 2004 "Habitat Predicts Losses of Red Grouse to Individual Hen Harriers". *Journal of Applied Ecology* 41: 305-314.

Bateson, G. 1979 *Mind and Nature: a Necessary Unity*. Cresskill: Hampton Press.

—. 1972 *Steps to an Ecology of Mind*. Chicago: University of Chicago Press.

Holloway, S. 1996 *The Historical Atlas of Breeding Birds in Britain and Ireland, 1875 - 1900*. London: T & A.D. Poyser.

Latour, B. 2004. *Politics of Nature: How to Bring the Sciences into Democracy*. London: Harvard University Press.

—. 1993 *We Have Never Been Modern* (trans.) C. Porter. Cambridge, MA: Harvard University Press.

Milton, K. 1996 *Environmentalism and Cultural Theory*. London: Routledge.

Pálsson, G. 1996 "Human-Environmental Relations: Orientalism, Paternalism and Communalism". In P. Descola & G. Pálsson (eds.), *Nature and Society: Anthropological Perspectives*. London: Routledge: 63-81.

Prestt, I. 1966 "Studies of Recent Changes in the Status of Some Birds of Prey and Fish-Feeding Birds in Britain". *The Journal of Applied Ecology* 3: 107-112.

Rappaport, R. 1971 "Ritual, Sanctity and Cybernetics". *American Anthropologist* 73: 59-76.

—. 1967 *Pigs for the Ancestors: Ritual in the Ecology of a New Guinea People*. New Haven: Yale University Press.

Ratcliffe, D.A. 1970 "Changes Attributable to Pesticides in Egg Breakage Frequency and Eggshell Thickness in Some British Birds". *The Journal of Applied Ecology* 7: 67-115.

Redpath, S.M., B.E. Arroyo, F.M. Leckie, P. Bacon, N. Bayfield, R.J. Gutiérrez & S.J. Thirgood. 2004 "Conservation in Practice. Using Decision Modeling with Stakeholders to Reduce Human-Wildlife Conflict: a Raptor-Grouse Case Study". *Conservation Biology* 18: 350-359.

Redpath, S.M., S.J. Thirgood & F.M. Leckie. 2001 "Does Supplementary Feeding Reduce Predation of Red Grouse by Hen Harriers?". *The Journal of Applied Ecology* 38: 1157-1168.

Royal Society for the Protection of Birds. 1998 "Hen Harriers - A Legitimate Grouse?". In *Birds: The Magazine of the RSPB*.

Scottish Gamekeepers Association. 2001 "Scottish Gamekeepers Association Response to the Scottish Executive Policy Statement 'The Nature of Scotland'": Scottish Gamekeepers Association.

—. n.d. The Raptor Problem - Our Solutions: Scottish Gamekeepers Association.

Smout, T.C. 2000 *Nature Contested: Environmental History in Scotland and Northern England since 1600*. Edinburgh: Edinburgh University Press.

Snow, D.W. & C.M. Perrins. 1998 *The Birds of the Western Palearctic: Concise Edition*. Oxford: Oxford University Press.

Strathern, M. 1992 *After Nature: English Kinship in the Late Twentieth Century*. Cambridge: Cambridge University Press.

The Game Conservancy Trust. 2003. "Hen Harriers and Grouse". The Game Conservancy Trust.

Thirgood, S., S. Redpath, I. Newton & P. Hudson. 2000 "Raptors and Red Grouse: Conservation Conflicts and Management Solutions". *Conservation Biology* 14: 95-104.

Thirgood, S., S.M. Redpath, S. Campbell & A. Smith. 2002 "Do Habitat Characteristics Influence Predation on Red Grouse?". *Journal of Applied Ecology* 39: 217-225.

Watson, M. & S. Thirgood. 2001 "Could Translocation Aid Hen Harrier Conservation in the UK?". *Animal Conservation* 4: 37-43.

Yearley, S. 1993 "Standing in for Nature: The Practicalities of Environmental Organizations' Use of Science". In K. Milton (ed.), *Environmentalism: The View from Anthropology* . London: Routledge: 57-70.

CHAPTER SIX

ANIMALS AND SCIENCE: BETWEEN TECHNOSCIENCE AND DEVELOPING RURAL AREAS IN NORTHERN ITALY

CRISTINA GRASSENI

Introduction[1]

The title of this chapter evokes a contradiction inherent in the agricultural landscape of northern Italy. This area is often associated with extensive rice fields, richly manured pasture lands and maize crops and is a world apart from a typical "developing" rural area. The Lombardy region, home to about ten million people or 16% of the national population (with two million of these living in the capital, Milan), is indeed highly industrialised. Lombardy is also responsible for slightly less than a quarter of the Italian national gross domestic product. Historically, the farming regions of Lombardy were the first in Italy to develop intensive and mechanised agriculture, and are now characterised by three intensive monocultures: corn, rice and pastures for dairy cows. However, not far from the intensively exploited plains lies the Alpine terrain. This area is heavily forested but also has Alpine pastures, both of which distinguish it from the lower plains. However, outmigration from the highlands has been a consistent feature here since the so-called "economic boom" of the Sixties, with many people moving from the Northern Italian Alps to the lowland industrial plants around Milan (specialising in steelworks, domestic appliances, tyres and chemical derivates) and in Piedmont and Turin (home to FIAT cars and related manufacturing companies).

More recently, lowland industries have fallen into deep crisis and experienced crippling decline. Simultaneously however, traditional farming and food cultures in the environmentally attractive hills and highlands of northern Italy have become once again a frontier for development. Since agribusiness can only be implemented fully in lowland terrains–both for geographical and environmental reasons–a current strategy for “highland rural development” is seeking to transform and reinvent subsistence agriculture. This is particularly the case in diary production: advanced reproductive technologies oriented towards "improving" local herds are being used to produce high quality milk from selected cattle. Such practices are focused locally on Brown cows of Swiss descent that are specialised in milk production.

The thesis underlying this chapter is that the role farm animals play in the local economy and in local knowledge illustrates how farms in Northern Italy are becoming sites of scientific and technological innovation (see Grasseni 2007). Farms are sites of mediation and negotiation between local farming knowledge and the technical advice of breed experts, as well as between legal regulations concerning milk-production and the local conditions and routines regulating the times and modes of milk-storage and cheese-making (see Grasseni 2006). In the case of contemporary food production, I propose that the borders of science and technology are increasingly difficult to determine. As the history of science (Shapin 2003; Schaffer and Shapin 1985) and the social studies of science have shown (Latour and Woolgar 1979), the boundaries of the laboratory are porous.[2] What counts as remaining inside the laboratory and in the domain of science, and that which resides outside of these can at times be difficult to determine. For instance consider bovine spongiform encephalopathy (BSE) or "mad cow" disease. BSE can arguably be understood as the result of an *in vivo* collective experiment, which as a result changed mainstream "timescapes" of risk perception regarding food (Adam 1998; Latour 2004). Of particular interest to me in this chapter is what happens when the laboratory "goes to the farm" and the results of scientific experiments become the object of cultural and political approval, rather than of an aseptic verification of hypotheses against experimental facts.

Thus, within such a framing, even remote mountain farms can be understood as sites of technological innovation. In this chapter, I will firstly demonstrate how farms in the Lombardy region are becoming sites of mediation and negotiation of local farming knowledge *vis-à-vis* technological advancements and the logistics introduced by breed selection practices. I then consider how regulations concerning milk-production, storage and cheese-making also influence the spaces, times and routines of the relationship between breeders and animals. Understanding this side of reproductive technologies means appreciating the social dimension, the “ecology of culture” so to speak, of a particular community of practice–that of milk cow breeders–which participates in breed selection. Finally, I examine a highly complex web of technical, political and social agencies that shape the practice and discourse of agricultural "improvement". This includes a wide spectrum of people who advise, monitor and assist mountain farmers and breeders, most notably vets, agricultural advisors and technicians from the agricultural trade unions, but also scientists, biotechnologists and statisticians involved in the selection and "genetic improvement" of progeny cattle.

Furthermore, extending beyond immediate on-site routines of farming practices are a nexus of economic pressures, worries over animal productivity and techno-scientific innovation that shape future possibilities of breeding. These include embryo-transfer, cloning, breed selection and "genetic salvage". Although these techniques are not the main topic of this chapter, I shall draw out here some of their general implications. What is their specific role for a new kind of political thinking about science and democracy in "developed" societies at large?

My argument is that in order to better understand farms as sites of technological innovation, and to explain their varying and sometimes contrasting degrees and modalities of "development", we need to explore the routines and webs of relationships that call forth the role of animals not only within the strict routines on site, but also within societies at large. Traditional approaches to the study of agricultural innovation (including science and technology, innovation systems, policy-making and governance) often focus exclusively on the logistics of farm services, on juridical regulations about food production, or on the distribution and targeting of subsidies. This study, however, considers that farms are complex contexts, determined also by the routines and relationships that they develop with both institutional and non-governmental actors, with both entrepreneurs and scientific innovators. As I hope this paper will show, animals and science in this context stand for complex constellations of actors perched between the frontiers of technoscience and the rigid economic ties to their contexts of origin–namely, in this case, “developing" rural areas.

Biotechnology and Dairy Farming in the Lowlands and Mountains of Lombardy

My fieldwork in the valleys of Lombardy began in 1998, when I made a film called "Those Who Don't Work Don't Make Love" as part of my training in visual anthropology in Manchester. In it, I followed one family of dairy farmers in the mountains near, in Lombardy, during their transition from traditional to modern farming, culminating in their construction of a new cattle shed with a milking parlour. To place the social and geographical characteristics of this region in context, one must bear in mind the following: Lombardy spans different terrains, from high altitude pasture lands to lowland areas devoted to commercial crops and intensive animal breeding. Furthermore, according to the most recent statistical surveys, it is also the region that spends the most in research and development in Italy[3]. At the same time, especially in provinces like Bergamo, the mountain areas have historically accommodated dairy farming which is not quantitatively competitive but which "scores well" through marketing typical or artisanal products, most notably mountain cheese. The valley I worked in for most of my ethnographic fieldwork, Val Brembana, is often depicted in political and socio-economic analyses of the Lombardy Region as "marginal", "rural" and "traditional". Like other similar mountain areas in northern Italy, agricultural aid in the form of European "structural funds" given to local administrations and distributed to working farmers amounts to thirty to forty percent of the income of all agricultural concerns. It is clear, then, that the economic viability of farming futures is tightly bound to the priorities of the Common Agricultural Policy and to the possible prospects of enhancing local cheese production through careful campaigns targeted at the market of niche and ethnic foods, as well as eco-tourism.

Dairy farming is still considered the "agricultural vocation" of the province of Bergamo, but in the last fifty years it has followed quite different development "paths" in lowland and mountain areas. In Val Brembana, the agricultural census of 1990 registered 3609 farms, of which only about 250 or 300 exceeded a size of three hectares. Of these 3609 farms, 1954 were breeding farms and 662 bred cattle, each with only nine cows on average. The number of dairy farmers registered for EU milking quotas in the year 2000 was about 200.[4] In 2006, 5,174 farms were registered in the whole of Bergamo province of which 1,577 were located on the plains, 1,829 around the town of Bergamo, and 1,768 in the valleys of Bergamo (386 in Val Brembana). One third of the territory is considered "a mountain agricultural system"–including the foot of the mountain terrain. Recent data (2007-2009) confirm the latest census, counting 1,768 out of 5,174 farms located within the mountain system, but with only 3% of the province population employed in agriculture.[5]

From these figures we can start to appreciate the social and ecological incidence of farming, despite its economic marginality. Many more local people keep animals, or own pastures or woodlands, than those who make a living out of farming. Even more remain hidden to statistics but have some engagement with "rural" activities, whether it is helping out relatives on their cattle farm or ensuring the provision of wood. Locating farmers on these terrains is important in the light of a discussion of their "marginality" and consequent eligibility to EU funding aid. The direction in which such funding, coupled with scientific innovation, takes local farmers is by no means uncontroversial. Being from Val Brembana is often a synonym for backwardness, social isolation and marginalisation, in the urban common discourse. For instance, "coming down from Val Brembana" is a common expression in Milan to denote lack of manners and uncouthness. This is both the effect of cultural stereotyping and a simplified image of a real and complex social picture. For instance, geographical isolation and lack of infrastructure may generate social and cultural fragility amongst farmers, especially for those who remain single, including frequent cases of alcoholism and depression that often result in the closure of small farms.

Nevertheless, the ratio of "young" farming entrepreneurs (under 40 years of age) to farmers over the age of 65 in the mountain area is 1.74, by far higher than that in the plains. This is an indicator of how the mountain agricultural system could be more open to social innovation and technological transfer, prompted by an intergenerational change of management. Within the community of practice of cattle breeders who invest in breed selection via artificial insemination (AI), the characterisation of the area as isolated and marginalised is still held to be true, and can be measured through the lack of availability of proper personnel and structures for implementing the AI techniques. As an illustration of this, in the early 1980s there were only a couple of lay inseminators in Val Brembana. Each typically performed six to seven hundred artificial inseminations of cows with the semen of progeny bulls annually, as requests for the procedure were high but the sheds were scattered and personnel scarce.

By contrast, the lowlands are a very different, but geographically very close, working environment. The production of bovine embryos of predetermined sex is described here as a fact of the “near future”.[6] The Laboratory of Reproduction Technologies (LTR), based in Cremona and founded in 1991, as a branch of CIZ Research and Genetics, is a consortium for animal genetic research and its commercial exploitation. It offers bovine, horse or buffalo embryos for embryo-transfer produced by its own originally developed "ovum pick up" techniques. The laboratory is located in the centre of the lowlands in the neighbouring province of Cremona, south of Milan–a region devoted to rice and corn monocultures and to intensive dairy farming. Here, thirty months after a cow first gives birth, her calf becomes a heifer. This heifer will in turn have been inseminated, will have given birth and will be lactating. At this point it becomes an economic imperative–according to a widespread practice in lowland intensive breeding–to maximise production by "substituting" the mothers with their daughters, since the latter carry updated and reliable "genetic material". In other words, the continuous striving for improving the herd's "milking traits" means that a cow's production is only exploited until the daughters are capable of a production that exceeds their mothers'. According to the creed of breed improvers, this is precisely the model that mountain farmers should strive to replicate, too. In contrast, it is precisely in the residual interstices of subsistence mountain farming, with no prospects of breed "improvement" or of firm "growth", that milking cows are still allowed to live out their natural productive cycle of up to 10 or 12 years.

Despite the fact that the Common Agricultural Policy aims to reduce the environmental impact of agribusiness with measures for the preservation and improvement of the rural landscape (as first stated in EU regulation 2078/92[7] and assumed as a strategic aim for rural development 2007-2013), the intensive production of the lowlands is still an economically more viable business than the environmental-friendly, low-impact traditional farming of the mountains. The recent fall in the price of milk however, and the rising concern about nitrogen pollution in soil and water due to intensive dairy farming, is belatedly opening up the debate on whether the overall results of breed selection constitutes an "improvement", or a liability.

Improving Animals

The drive toward selective breeding has characterised the history of animal husbandry in Europe and beyond.[8] The introduction of breed selection by inbreeding plays an important part in the political history of "improvement". In the Lombardy mountains and plains, this phenomenon has profoundly changed what was a well established breeding practice, connecting highlands and lowlands through seasonal circuits of transhumance. Transhumant herders from the Lombardy mountains have taken their cattle to the lowlands to graze during the winter since the fifteenth century. This implied a continuous blurring of social roles whereby mountain breeders could become cheese merchants or lowland loaners and owners. With the increase of irrigated lands for crops, and the intensive production of grass for pasture, the sedentarisation of cattle interrupted an osmotic relationship with the highlands–a well established phenomenon already in the nineteenth century. In Britain, the enclosures movement, the rise of new fodder crops, a generation of agricultural pioneers and popularisers together with a number of institutions and societies established in the last two decades of the eighteenth century prepared the ground for the import of choice livestock. Selective inbreeding then produced the first thoroughbred breeds, racehorses, then beef cattle, pigs and sheep (Fussell 1984: 26-45). As for agricultural improvement in continental Europe, Barbara Orland (2003) stresses the link between treating soil, animal and fodder as an economic and ecological unit (from the end of the eighteenth century) and the consequent pressure for competitive dairy cattle in the nineteenth and twentieth centuries (see Grasseni 2007, 2009).

Various breeds of dairy cows, including the Alpine Brown, developed locally in different valleys of the Alpine region. Specimens of the Swiss Brown breed were exported from Switzerland to the U.S. and selected as a milk-producing breed in the 1880s, only to be virtually re-imported to Europe after World War II in the form of the semen of selected "progeny" bulls. The virtual erasure of national cattle stocks was followed by a consistent reintroduction of selected strands of "improved" breeds such as the US Swiss Brown. As a result, several local breeds were lost whilst improvement by inbreeding took place on a massive scale. Nowadays, according to the US Brown Swiss web site almost all Brown cows today are descended from an American cow, Jane of Vernon, Grand Champion in 1932, 1933, 1934 and 1936.[9]

The same philosophy of improvement is applied today, aided by vast statistical and empirical databases about milk production and fertility. For instance the "Superbrown" consortium gathers and tests bulls from the area of Trento and Bolzano (Trentino Alto-Adige). Details of their genetic selection programme are on-line: http://www.superbrown.it/ . This is an artificially selected breed yielding more milk than the "traditional" Alpine Brown which is now virtually extinct in Italy, although a strand of the original breed is maintained in Switzerland. In 1997, forty years after it was established, it was understandably a source of pride for the Italian National Association of the Brown Breed (Associazione Nazionale Allevatori Razza Bruna or ANARB) to be able to export semen for the first time from Italian Brown sires to the US.

Having sketched, however briefly, some of the most important macroscopic traits of the context I have studied–namely, the state of the art and the historical roots of dairy breeding in this region–I now wish to consider in more detail the ethnographic fieldwork I carried out with a family of dairy-cattle breeders in the Bergamasque Alps. In particular, I shall highlight some of the technical and political ways in which the daily routine of breeders, breed inspectors and agricultural consultants incorporate "reproductive knowledge".

Logistics and Kinship: Living as a Dairy Breeder in the Italian Alps

The family I lived with practices the *alpeggio* (French: *alpage*) also known as *monticazione* - taking the herds annually to higher pastures to graze the grass during the hottest weeks of the year.[10]

Living on the pastures with my host family meant having several opportunities to meet fellow breeders with neighbouring herds, to visit clients from the lowland whose herds were being tended by "my" family, and to talk to the vet and the various breed experts and agricultural technicians who visited the herd and inspected the milking and cheese-making structures. As a "resident anthropologist", I witnessed and filmed various operations such as assisting cows giving birth, suckling calves, tooth-extractions, incisions, hoof-trimming, soling and repairing, and of course artificially inseminating heifers. In the case of the latter, I was given detailed explanations of the criteria for choosing bull semen, and I accompanied a member of the family who, being a licensed lay inseminator, offered his services to neighbouring herds.

The most interesting aspect of my stay in the "upper pastures"–given the stereotyping of bergamasque farmers as backward, antisocial and isolated–was the complete *lack* of isolation that it implied. On the contrary, every day was a new social adventure. Apart from the five practising professionals of the family, together with their wives and children, I met one vet, five operators of the local "service centre for agriculture", the head of the local agricultural trade union bureau and that of the provincial bureau, one freelance agricultural consultant, two visiting breeders from the lowlands, one lay inseminator working mainly with Friesian herds, four neighbouring herd-keepers and their families, and two technicians of the local breeders' association section. Keeping a systematic list did not make much sense at the time since the rationale for my presence was not to gather quantitative information nor to administer questionnaires to professionals, but to absorb and later reconstruct the complex web of relationships, both of friendship and of professional exchange (the former often existing *because* of the latter) and to witness the conversations and practices surrounding my hosts. I did not have to steer the conversation toward the main issues of their profession (the best bulls and cows available in the market, the latest winner at the local cattle fair, the problem with milk quality checks and the rise in somatic cell count, the price of milk, the political allegiances and fights between personalities in this or that local or regional office of some association or other...) since these were the topic of endless chat and gossip between the various people I encountered.

Later on, with ANARB breed experts I observed the evaluation of cattle in local, regional and national fairs and received filmed evaluations of prize cattle at international fairs. During fieldwork, I visited the national fair of the Italian Brown Breed in Verona (February 1999), the inter-regional agricultural fair of the North at Montichiari, including both Friesian & Brown breeds (March 1999) and the local cattle fair of Valtaleggio at Sottochiesa (October 1999). I accompanied one breed expert on his checks of all the registered sheds of Val Sabbia, in the neighbouring province of Brescia, and I joined a breed inspector on his check of dairy farmers in Valtaleggio (twenty-two in all) and thirty-eight registered members of ANARB in the wider basin of Val Brembana which encompasses Vataleggio. In Valtaleggio I also visited all the minor farms keeping sheep and goats in the company of the local vet on his TB checks. With the trade union technicians of the Val Brembana office I had the chance of visiting small-scale farmers who were introducing cheese-making and ecotourism as part of their farming practices and visited the two existing cooperative sites for milk-storage and cheese production, one in Valtaleggio and one in Upper Val Brembana.

Such grounded engagement with trade union officers, local breeders and operators of the breeders' association, as well as my conspicuous presence over two years on a relatively small territory where everyone knows everyone else, especially in the trade, allowed me to widen the circle of acquaintances and also to come into contact with and follow professionals who came from the mountain community but who now work on lowland farms. This includes breeders, lay inseminators and agricultural consultants. Between 1999 and 2001, I was able to visit intensive breeding farms in the lowlands just south of Bergamo and about sixty kilometres from the valley. This offered an opportunity to observe the completely different logistical, geographical and structural environments in which dairy cows, especially Friesians, are bred.

Often I witnessed the ascent or decline of the political and professional careers of local officers whom I had met as young graduates. As for the fortunes of my hosts, yearly visits to the high pastures and to the village stable have allowed me to monitor the growth of the herd and their rising notoriety in the valley (at the local cattle fair in October 2004, 2005 and 2006 they won virtually all the prizes) as well as the professional careers of the family members in their progression from provincial to national breed experts, and the evolution of their relations with high-ranking officers of the breeding associations with which they are involved. Indeed, each member of the farming family that hosted me (mother, father, four married sons and their offspring, totalling 16 people) is involved in the family business, which has invested greatly in the farm's scale and productivity. Each of the three married brothers working on the farm has developed a breed-related specialist competence, and a fourth works as a cheese-maker. The youngest and eldest sons are national ANARB experts, which means that they may act as judges in cattle fair competitions and as evaluators of selected herds. As part of his professional updating, the eldest son has visited the U.S. Brown Swiss association headquarters and farms in Wisconsin and is very proud of it.

None of the wives are employed outside the farm, although two are employed part time in the local cheese-making cooperative. They are busy keeping house, haymaking in the summer, keeping chickens, maintaining a garden patch, and selling their home-made cheese to visitors.

The only male grandchild, now a teenager, is fully involved in the running of the farm and has been apprenticed to his grandfather since a very early age, tending the herd on the high pastures during the summer school break, taking part in milking, removing manure, feeding calves, and so on. As one may imagine, each member either participates wholeheartedly in the common enterprise or is bound to adopt a position progressively further "at the fringes", as is the case of the only married son who does not live on the premises and who has not formally joined the family company.

Three tractors (for transport to and from the high pastures, haymaking and operating the mixing cart) and all the necessary tools to handle mechanised haymaking (grass-cutting, turning and dispersing the grass, drying in a ventilated chamber and wrapping in plastic bales) make an impressive battery of machinery, which is by far the largest in the valley. In 1981 the family cattle-shed was enlarged for a first time, from the traditional *stalletta* to a proper stable. This building now hosts only dry cows and heifers, whilst an adjacent room (the initial family cowshed) hosts only weaned calves. Originally, cheese-making also took place in here. While this would still physically be possible, it would be now illegal. In the 1981-built stable, cows were still chained to their posts, and there was a central corridor to distribute fodder together with rear gutters for manure and individual water buckets. The problem with this type of "traditional" stable is that it is not suitable for "selected" "progeny" cows which are much bigger animals than their predecessors and end up standing with their rear legs in the gutter at the back of their post, risking hoof infection and soiling of the udders. Moreover, through being chained to the spot throughout the winter months, the cows' muscle tone becomes reduced and some become incapable of getting up, while others develop walking difficulties.

The current modern stable was built during my first fieldwork in 1998 and houses 40 to 50 cows that are free to move around, can access food and water in allocated areas, and walk directly into the milking parlour. Stimulation for milk-yielding is mechanical inside the sucking cylinders, so that in theory no contact with human hands is necessary.[11] The milk is conferred on the local dairy cooperative for cheese making, although homemade cheese is also sold. In fact, during the summer, milk is transformed and sold on the premises of the high pastures, or sent to the village homestead, for maturing.

The stations around which life is organised during the eighty days of the *alpage* include lodgings and cowsheds with a generator to allow machine-milking. Ideally, the first station serves as a transition phase: the cows graze on grass during the day but are sheltered in a shed at night and are given hay and fodder. In my hosts' herd, the most productive cows are aided with concentrated fodder and salts even on the high pastures. The lodgings, originally stone-huts, have been restored and provided with a cellar for cheese-making and maturing. At the time of fieldwork the first station (at 1,400 metres) was only reachable by tractor or Land Rover. Subsequently, the municipality has made the first and second station (at 1,700 metres) more easily accessible to four-wheel-drive vehicles. Summer grazing in the upper pastures is healthy for the cows and very economical for the family, which can save up to a quarter of the yearly need for fodder, and also make hay in particularly rich fields. *Alpage* cheese is a commercially viable production, thanks also to connections with the Slow Food movement for the preservation of "authentic" rural food production.[12]

The practice of *alpage* also has disadvantages, which include the risks of taking livestock up mountain trails and exposing them to the elements, for a period of up to four months, in conditions where no vet is readily available and storms are likely to strike. Having a cow struck by lightning or falling into a precipice means losing both production and bloodline, plus the cost of calling a vet to the high pastures at one's own expense to report the death and have it certified. The animal cannot be butchered for meat and the carcass must be disposed of as instructed by the vet, using up precious family labour that could be employed in haymaking at the village or in the pastures. Minor injuries such as hoof-damage or dislocated shoulders are also regular occurrences. Moreover, milk production drops dramatically on the pastures, both because of climatic change (from a sheltered and static life in the shed to exercise and exposure to sun, rain and cold), and because of the dietary change from "unifeed" mixes to grass. The advantages outweigh the disadvantages, in economic terms, only if the cheese produced in the high pastures is sold for a high price. The cultural and environmental significance of this tradition is also recognised by EU funded projects that aim at restoring and encouraging the practice of summer grazing.

Despite their continuing practice of the *alpage*, my hosts were amongst the first who wholeheartedly adopted AI for breed improvement in the 1970s. They are now proud that their productivity per cow has practically doubled (they also own a herd about ten times the average size of those of other farmers in the same valley). They consider themselves to be all-out *innovators* in their professional practice and in their social context. For instance, the advent of a new road network that allows easier access to the lower reaches of the upper pastures by tractor has meant that they readily abandoned the practice of Spring and Autumn grazing on these medium height mountain reaches (at about 1,200 metres). Instead, in about 2002 they started to make hay on these pastures with their updated machinery. The grass thus reaches the cattle in their village stable in the form of hay, rather than having to take the cattle up to the pastures in late Spring to eat the grass. The cows also eat in the Autumn the hay that used to be cut during the Summer with manual implements and stored on location, before road access was possible.

This eager transition to an intense year-round productive regime is also apparent in their management of the animals' diets. Their most productive cows are given an extra dose of calcium and glucose at the peak of lactation. This can also be administered by syringe if a cow weakens after giving birth. Suckling calves are given milk and hay but are not allowed to suckle from the mother. Weaning calves are given whey, the by-product of cheese, while the cows' diet is mostly solid. The family also took advantage of an experimental service called "genetic salvage" offered by the Laboratory of Reproductive Technology (LTR, see above). In 2005, when their best producer had to be put down, five eggs were extracted from the cow by "ovum pick up" techniques and, of the five ova, four produced suitable embryos of which two survived when implanted. One of the two (male) calves was taken in by the "genetic centre" of the national Brown breed association for progeny testing, the other by a competing consortium.

Figure 6-1: A tractor carrying fodder to the stable, Valtaleggio, 2004
(Photograph: C. Grasseni)

The most obvious consequences of the introduction of selective breeding in mountain environments, in terms of ecological sustainability, are that the increased size of the animals forces farmers to provide bigger pens as well as more and more diversified fodder. This involves seeking out a nutritionist's advice, providing bigger portions, and introducing maize and soya plus mineral and vitamin integrators into the animals' diet.

Such modifications in turn create a previously inexistent problem: that of disposing of liquid dung. This is usually spread onto local pastures, which may cause pollution of groundwater and causes a change in pasture vegetation due to increased amounts of nitrogen. This was demonstrated by a comparative study of two dairy farms in Valtaleggio, one with more than one hundred and fifty cows and one with fifteen, with reference to their different management of the pastures and related dangers to soil and water-pollution.[13]

The transition from a seasonal regime to an all-year intense production regime mentioned previously also means that the *alpage* has less and less significance in the way in which the farm's management is conceived.

Labour division tends to be "rationalised" in the sense of trying to avoid having labourers on the high pastures when they could be employed in hay making or in tending cows in the village stable. Overall, these three factors combined have obvious results not only in terms of environmental sustainability but in terms of social and cultural conservation.

Despite the current environmental alarm concerning the Lombardy lowlands where the soil and water are already polluted with nitrogen-based substances derived from the accumulation of manure in the area, intensive farming continues to be the only model applied there and is also currently promoted as a viable model for the mountains. Whilst professional associations and trade unions assist and prize “innovative” and “technologically intensive” practices, local operators will not be dissuaded from intensifying their farming methods. For instance, my hosts will not be convinced that they are doing anything wrong while they keep reaping all the prizes at the local cattle fair, which they consider a sure sign of their superiority and an official acknowledgement of their professionalism.

The Social Costs of Innovation

My hosts’ “social advancement” by force of technology is creating obvious rifts within the community, as I was able to observe over a longer period of time. Ten years down the line from their decision to modernise, the consequences of their innovations on the community farming economy are apparent. Some observations are worthy of attention, as they help us to assess the *social* sustainability (or rather as I am arguing, the *un*-sustainability) of the introduction of competitive and intensive farming in small mountain communities. Firstly the prominence of kinship ties was made obvious at the valley’s cattle fair, an event which in fact provided a stage for a performance by the whole family as well as for their cows. At the 2004 event, the leading farming family in the community put itself on display with the adults gracefully leading their best exemplars and the children also participating by leading calves and young steers.

The family also implicitly boasted of its economic success by bringing along a foreign salaried worker who cleaned up the cows on behalf of family members before the latter led them in the ring. They won so many prizes that they eventually had to cart them away in a wheel-barrow (itself one of the prizes they had won). This excessive display of success sparked a fair amount of verbal animosity and envy as testified by the looks of the audience and fellow farmers.

Figure 6-2: Cows and men lining up by categories at the local valley's fair, in 2004 in Valtaleggio (BG). (Photograph: C. Grasseni)

Most other breeders of cattle have been out-sized, have retired or are selling off. Nevertheless, during the cattle fair in the valley in October 2004, I observed that the event had retained its character of social relevance despite the fact the most families in the valley no longer actively farm. This is indeed a case in which a whole community of farmers and ex-farmers has come to identify with standards and criteria of excellence in breeding that came from outside. Such criteria impose conditions for success that in fact only a privileged few of them–one family of emerging entrepreneurs who have successfully invested in breed selection and artificial insemination with progeny bulls–can meet.

Figure 6-3: Local breeders who have won no prizes intently look on while the prize-giving ceremony takes place at the cattle fair. Valtaleggio, 2004 (Photograph: C. Grasseni)

We can then provisionally conclude that the introduction of technoscientific innovations in this one family's management of their cattle, and of the division of labour for the *alpage* season, has had a momentous impact on the collective social and cultural environment of the locality. It was in fact a collective experimentation, but one without a foreseen complex scenario that anticipated the ecological, cultural and social impact of innovating animal husbandry. It appears clear that this model for agricultural development and animal "specialisation" or "improvement" raises the issue of the environmental sustainability of the use of reproductive knowledge for intensive farming in mountainous terrains, as well as the further issue of its social and cultural sustainability as a model for local development. On the other hand, new social actors, such as the Slow Food association, are increasingly concerned with the quality and the sustainability of food production. They are proposing alternative models of farming and of food production, and beginning to experiment on a local basis. In Val Brembana, in the course of 2009 a dedicated and forward thinking Slow Food convivium has mapped existent "traditional" producers of *stracchino* cheese, in order to safeguard sound practices of animal breeding and of artisanal cheese production. They are establishing a Slow Food presidium which will guarantee that such cheese is made of raw milk and no additives are used in the cow's diet. These kinds of initiatives are ground-breaking since they provide farmers with new interlocutors who apply different criteria to their practices: a very small scale farm, family-based production and a herd sized 5 to 50 cows is not considered, in this case, residual or marginal but rather "traditional patrimony" and "sustainable best practice". The following sections will help put this provisional conclusion in a wider framework concerning the role of experimentation with animals and the life sciences in this particular socio-economic context.

The Social Appropriation of Technology

Many ambivalences, negotiations and compromises coexist in the practice of applying "reproductive knowledge" to breed selection. For instance, ranking specimens according to their genealogy, assessing the statistical inheritability of individual traits, or weighing each factor that adds up to the "total economic index" of a reproducer (from "milking speed" to the "protein count" in milk) can all often be *political* and strategic decisions, which may distinguish one breed association from another. In other words, genetic indexes are subject to economic negotiations and historical evolution.

Even though I did not witness it in person, the encounter between my host family and the scientists of the Laboratory of Reproductive Technology in the lowlands of Cremona marks for me a significant turn in my ethnographic understanding of the similarities, differences and mingling between "mountain" and "lowland" practices of breeding. It also sparked the beginning of a line of enquiry in my own work which is still largely "in progress": a "natural" curiosity for the topics of specialised biotechnological research connected with animal husbandry, in particular mammal cloning. My own understanding of the community of practice of breeders of the Brown breed in Italy motivated me to try to intuit at least, the significance of frontier biotechnical research for their own practices and prospects. So far this curiosity has not resulted in an ethnography but rather in a number of visits to LTR and an extensive interview published on line on the web site of the Milanese foundation "Fondazione Giannino Bassetti" for responsible innovation.[14]

Selective breeders make widespread use of AI, a low-impact technology, which nevertheless would not be available to them on their terrain without considerable investment in logistics (semen storage in liquid nitrogen, transport and distribution) and breed selection (testing of the progeny bulls in ad hoc "genetic centres", collection and processing of data of the milking performance of the offspring, regular publication of databases and magazines with updated information about best performers, monitoring of the producers in the sheds of all registered farmers nationwide). Technoscientific developments then make higher-impact techniques available for maximising the results of breed selection. For instance, consider embryo-transfer which consists of both pharmacological and veterinary interventions on the best reproducers to obtain a high number of genetically identical embryos, to be implanted in as many heifers.

Outside the circle of biotechnologists, vets and selective breeders, few know that deep in the Lombardy plains, not far from Milan, cutting-edge international research has been conducted since 1991 by professors Cesare Galli and Giovanna Lazzari in the Laboratory of Reproductive Technologies (LTR) which until 2008 was a branch of the Consortium for Zootechnology (CIZ) of the Italian Breeders Association (AIA). Starting from original protocols for the production of gametes and embryos of breeding animals (bovines, horses and pigs), this team has successfully developed a technique for mammal cloning through somatic cell nuclear transfer (SCNT)–the technique that gained publicity through its association with the birth of Dolly the sheep (see Neresini 2000, Franklin 2007).

The term "techno-science" which I apply to this context of research is apt, I think, to denote the growing proximity of basic and applied research[15]. This is probably the first time in history that socio-technical networks (to borrow a phrase from Bruno Latour) "know" and "can" in equal proportion and in highly interconnected ways. This is clearly the case in the field of research into animal husbandry and in particular in biotechnological applications to animal reproduction. Both ethical and political scenarios are hence called forth by this novel degree of proximity. Debating these scenarios means posing a question, not so much of the ethics of research or of the social responsibility of the researcher, but rather of the responsibility for the results that ensue from innovation. Such results are not only brought about by researchers but by their implementation in the farming economic and ecological system. If innovation is "the realization of the improbable"[16], in the case of LTR, the novel degree of proximity between knowledge and power in the realm of animal biotechnology has sparked a "tacit controversy" about where this should lead us.

By "tacit" I mean that the question of how to use and implement technoscientific innovation is largely un-debated in the public sphere, whereas the direction that research should take is often a highly controversial object of media attention, sometimes leading to harsh decisions in the realms of research funding and of research institutional strategies. One example of such a decision, and of the lack of public debate about it, comes from the web site of the animal rights association *Beyond species (Oltre la specie)* which accuses the scientists of LTR of hypocrisy since they claim to be working to eliminate the use of laboratory animals (as some of the techniques associated with embryonic stem cells could lead to the development of systematic *in vitro* testing on such cells rather than on animals), but in order to carry out their programmes, they inflict artificial lesions on their own cloned pigs in order to then experimentally cure them with stem cell therapy[17]. The eagerness of farmers to apply technoscientific innovation to their own farmsites alerts us to the fact that here, animals do not raise further or different considerations that those raised by crops, such as in debates over GM foods.

On the other hand, during the interview I conducted with Cesare Galli and Giovanna Lazzari, the cultural contextualisation of the moral significance of their research emerges clearly. As stated by Cesare Galli himself:

> "It's a question of values, or rather, of justification. We don't justify cloning for the sake of producing food, but we do justify it for the sake of curing our own diseases (i.e., with xenotransplantation). In China they would not put it this way"[18].

In fact, it would appear that the availability of resources for research is a direct measure of the approval that such research meets by public administrators and decision-makers. This may appear to be a rather banal observation, but it is in fact the basis of concerted studies both in the history and the sociology of science.[19] The LTR scientists are themselves aware of this fact when they express their concern that scientific research should be explained and communicated "correctly" "in order to create a sensible understanding of the motivations and aims of scientific activity, by the public opinion". They also complain that politicians and decision-makers, themselves science-illiterate, "often gain an opinion about science not from specialised publications or from the experts but from the media, in order to base their decisions and orientations about scientific activities" which in turn exposes them to "the environmentalist, the animal rights and the religious" lobbies (Galli 2003).

The case of LTR is emblematic in many ways of this entanglement of science, politics and popular opinion: when the laboratory obtained the first cloned progeny bull in the world (Galileo, of the Italian Brown Breed), the bull was seized by the police as a result of the local magistrate's enforcement of the government ordinance of March 1997 which prohibited mammal cloning in Italy *including* for experimental purposes.[20] This climate, which effectively continued to inhibit research by prohibiting its commercial exploitation, eventually led to the closing down of LTR as a branch of the Italian Breeding Association on the basis that this organisation is not interested in investing in basic research if it is unable to exploit the bull semen and other commercial items (such as clones of racing horses etc.) that can be obtained as a result.

Understandably, Galli's line of argument would point towards a "technocratic solution" (Bucchi 2006) which combines a good campaign for the "public understanding of science" and the effective delegation of all decisions about science-making to the scientists themselves. The limits of the technocratic solutions are the fact that often decisions about the directions and limits of scientific practices are *political* and have to be taken within complex scenarios which do not only include techno-scientific considerations. It must also be taken in consideration that a high degree of scientific literacy does not always correspond to a higher degree of trust in science and scientists.[21] This dilemma has been presented clearly by a working group of social scientists in a report to the European Commission which is very critical of a technocratic solution to questions of technoscience, including biotechnology.[22]

This controversy seems to take us well away from the alpine pastures of Lombardy, but it proves how close the global problem of the *governance* of science–and of animals in science–is to the actual practices even of small mountain cow breeders.[23] The scientists working in the laboratory, lost in the fogs of the Lombardy plains, actually deal on a daily basis with prize horse-breeders and cattle-breeders seeking their services of embryo-stock production, sexing and "genetic salvage". These scientists are thus not only researchers, but also science communicators, politicians, and administrators, while the breeder is also a customer in a global network of products and services which intimately changes and determines animal welfare, health and lifespan. This direct-to-customer relationship actually impacts on a much wider ecological and cultural setup, such as that of the *alpage*, which concerns animal lifecycles as much as human lifestyles. In perspective, this intimate feedback loop poses a problem, not only of the governance of technoscience, but more widely, of collective modalities of choice and of involvement of the relevant stakeholders in the scenarios of innovation.

Conclusion

This chapter draws upon the contrasts between different interpretations of what "innovation" may mean in the realm of animal husbandry within the varying contexts of the mountains and of the lowlands of Lombardy, Northern Italy. One way of perceiving the stark contrast between mountain and lowland agriculture is through a comparison of the intensity and speed with which biotechnological practices–initially artificial insemination, later embryo-transfer and now experimental mammal cloning–have been introduced in the respective farming, cultural and entrepreneurial environments. An assumption prevails that the mountain region has a lesser capacity for "catching up" with biotechnology in the lowlands of the same region. This lesser capacity is usually ascribed to the supposed social marginality, cultural backwardness, and logistical difficulties of mountain agriculture and animal husbandry. But a closer look shows instead that an osmotic relationship between the highlands and the lowland, historically testified to by the tradition of long-distance transhumance, is still in place, though deeply transformed.

A glimpse into research laboratories (significantly placed right in the middle of the productive lowland intensive agricultural district) demonstrates that new research on mammal cloning for breeding purposes helps to enlarge the picture by allowing us to refocus on the role of animals within these contexts. Contrary to stereotypical assumptions, mountain farmers do indeed have access to the most recent development in biotechnology. Furthermore, this permeates their daily practice in multiple ways from reproductive techniques for their herds to their daily milking practices that employ self-auditing protocols such as HACCP, which were initially introduced by NASA engineers to analyse "hazard critical control points" in the manufacturing and storing of food for manned missions into space. Technoscientific interventions like HACCP food safety monitoring protocols are entangled with a "politics of nature" (Latour 2004) that make up the very fabric of the global food provisioning system, and takes us much further, as is evident in Val Brembana, to mammal cloning and frontier biotech research. This chapter has sought to relocate farm cattle at the core of the manifold significances and functions of reproductive technologies employed in the regions' farms and in the lab as sites of techno-scientific innovation, where every decision making process is a mix of rationality, contingence, power, and compromise (Hagendijk and Kallerud 2003). Ultimately, this is an attempt to examine animal husbandry practices in detail, and to read some of their most problematic aspects through the lenses of the debate about the role of nature in socio-economic development (Dickens 1996; Escobar 1999; Hobart 1993) and about the governance of technoscientific innovation (Pellegrini 2008).[24]

As we have seen from ethnographic evidence, at the valley's cattle fair as well as in the laboratory for reproductive technology, animals, social actors and technoscientific innovation intersect in a grey area that equally involves technology, politics and ecology. The introduction of biotechnological innovation as part of the standard practices of mountain cattle breeding changes local social interactions and has had permanent consequences. The rural is only apparently remote and secluded. It is in fact enmeshed in a constant exchange of "technical" advice, monitoring visits, expert evaluation, audit data-gathering, bull semen, sterilising agents, cultivated bacteria for fermenting the cheese-maker's curd, and so forth. As a result, breeding knowledge emerges from a network of multiple sites: the cattle fairs, the farms, the labs and the summer pastures. It is the genetic materials from the lab that allow the maintenance of a chain of high-standard breeding practice on the alpine pastures with frozen bull semen from selected progeny cattle being provided in nitrogen containers. It is the expert eye of the cattle fair judges that selects and evaluates the offspring of such progeny bulls, which will be employed all year round in the farms for milking. It is the economic result of such productive practice that is displayed with pride in cattle fairs, though it goes under the name of "genetic improvement". When such economic success is threatened, and a progeny cow is ill or dead, it is the "magic" of the most advanced biotechnological techniques of "genetic salvage" - cultivated at LTR - that allows the retrieval and insemination of the last eggs of the lost progenitor, and puts the "genetic" investment in her to work with new born calves.

Such interlacing networks and practices raise many questions as to the role of animals in Western industrialized societies and about the ethical premises and implications of the use of bred animals as carriers of "genetic" material. I have only been able to begin to outline these issues in this chapter, and further writing and research would be valuable in order to fully delve into them, opening as they do a Pandora's Box of interdisciplinary questions. What should be clear, though, is that beyond the stark apparent contrast between small-scale mountain agriculture, assumed by outsiders to be backward and isolated, and an advanced industrialised intensive agricultural practice in the lowlands, the rural alpine world is compromised and enmeshed in a network that tightly connects it to the sites of production of scientific knowledge and of global market rules. Laboratory techniques at the forefront of biotechnological innovation become pertinent and precious tools as increasing economic competition pushes alpine dairy farmers to make the most of their animals as "genetic" investments.

Notes

[1] Special thanks to Maggie Bolton and Cathrine Degnen for their scrupulous and painstaking editorial work.

[2] As Bruno Latour convincingly showed in his milestone study on the "pasteurisation of France" (1988), Pasteur convinced the scientific public of the existence of microbes, and then of the efficacy of his own vaccination technique by preventive inoculation, only *after* he found political and media allies amongst the Breeders Associations and in the public opinion itself.

[3] On average, one fifth of such spending is from private initiatives whilst the rest comes from public funding. (See *Annuario Scienza e Società 2009*. Observa - Science in Society: 90-91).

[4] Comunità Montana Val Brembana, *Piano di Sviluppo Socio-Economico* (2000: 29); *Piano Agricolo Provinciale 2007-2009*, http://www.provincia.bergamo.it/ProvBgSettori/provBgSettoriHomePageProcess.jsp?page=&myAction=&folderID=54159&notiziaID=92274. Last visited February 28th, 2010.

[5] Provincia di Bergamo. *Piano Agricolo Forestale 2007/2008*, pp. 7-8. Online at http://www.provincia.bergamo.it/provpordocs/Capitolo%202%20-%20Introduzione.pdf last visited on October 25th 2009.

[6] See the LTR website www.ciz.it.

[7] This included the reduction of the pressure of bovine herds on the pastures (measured in number of adult bovine units per unit of grazing land), the re-introduction of species or breeds at risk of extinction, and the re-conversion of arable land to fallow fields (as well as re-introducing hedge-rows and investing on reforestation in the lowlands). The Goteborg European Council Meeting of 2001 has confirmed that sustainability is a common aim for rural development.

[8] The many links between the theories of agricultural "improvement" with the European colonial enterprise have been focussed upon recently (see Miller and Reill (1996) and Drayton (2000) on the role of the "Banksian empire").

[9] The herd book of the Brown Swiss Association was established in Wisconsin in 1881 with the import of 25 bulls and 130 heifers from Switzerland (Eugster 1997: 24). Artificial insemination of autochthonous Swiss Brown cows with American Brown-Swiss bulls was authorised by the Swiss Federation of Brown Cattle breeders in 1967.

[10] A terminological specification may not go amiss here: Paolo Sibilla (1991) following the historian Charles Parain, distinguishes between transhumance and *alpage* as the former implies long-distance movements in any season, both in the lowlands and in the mountains, and is usually practiced by nomadic breeders, often with ovine flocks. *Alpage* instead is a form of seasonal grazing, practised by dairy farmers that for the rest of the year live stably in their mountain villages. Schuyler Jones (2005) calls instead "transhumance" exactly what the previous authors

define as *alpage* (namely, the seasonal movements of livestock from permanent villages practising agriculture), in order to distinguish it from "pastoral nomadism".

[11] Manual stimulation can spread mastitis by contamination through the hands.

[12] There is no space here for elaborating on the Slow Food movement and their campaign for the preservation of "authentic" enclosures where original recipes and foods (especially cheese) are supported through media and marketing campaigns. Slow Food Presidia are projects for the conservation of "endangered" foods and of the breeds involved in their production (http://www.presidislowfood.it/). Conserving and valuing traditional systems of production, recipes and nutritional standards for local animals is a substantial part of the effort of Slow Food Presidia. According to the Slow Food Foundation for Biodiversity (n.d.) there exist about 200 such presidia in Italy and another 120 worldwide. As I have elaborated elsewhere (Grasseni, in press) the modernisation of mountain farming of the sort carried out by my host family is in tension with the expectations of conservation of Slow Food Presidia which insist, for instance, on a fodder-free diet for dairy cows.

[13] *Il sistema zootecnico territoriale della Val Taleggio: aspetti produttivi, storici e socioeconomici*. Tesi di laurea di Cristina M. Bianchi, 2002/2003.

[14] *La ricerca di punta biotecnologica in Italia è una chimera? Una intervista a Giovanna Lazzari e Cesare Galli*, http://www.fondazionebassetti.org/it/grasseni/2008/10/la_ricerca_di_punta_biotec nolo.html

[15] Massimiano Bucchi, *Scegliere il mondo che vogliamo*, Bologna, Il Mulino, 2006, p. 9.

[16] Innovation, as defined by Piero Bassetti, is "the realization of the improbable", namely the actual happening of an improbable fact thanks to the joining up of novel knowledge and the power to make it real. Innovation is hence both risk and opportunity, something that changed the world in intrinsically unforeseeable ways. The unforeseen can happen on the social-political level (new institutions, new models of relationship, production, war, new powers), on the technical-economic level (new materials, new energies, tools, or types of commodity), or on the cultural-aesthetic level (new styles, fashions, tastes and attitudes) (Piero Bassetti, *L'innovazione responsabile*, in AA.VV. *Innovazione, creatività, responsabilità. Formare gli imprenditori del futuro*, «Quaderni della Fondazione Giannino Bassetti», 1/2006, p. 20).

[17] http://www.oltrelaspecie.org/nemici-scienza4.htm

[18] *La ricerca di punta biotecnologica in Italia è una chimera? Una intervista a Giovanna Lazzari e Cesare Galli*, http://www.fondazionebassetti.org/it/grasseni/2008/10/la_ricerca_di_punta_biotec nolo.html

[19] See for instance the work of Peter Galison such as his film *The Ultimate Weapon: The H-Bomb Dilemma*, (Galison 2000), and of Helga Nowotny (e.g. Nowotny *et a.l* 2001).

[20] The ordinance was itself "cloned" several times, in that it had to be reiterated periodically as a governmental, rather than a parliamentary act, to ensure its

reinforcement, as notices ironically the philosopher of law Mariachiara Tallacchini (2002).

[21] Eurobarometer 2005, quoted in Bucchi (2006: 11-46); M. Bucchi e F. Neresini, *Gli Italiani sono contrari alle biotecnologie? Alcuni stereotipi su opinione pubblica e biotecnologie in Italia*, in Bucchi e Neresini, *Cellule e Cittadini. Biotecnologie nello spazio pubblico*, Milano, Sironi Editore, 2006.

[22] See the report signed by Ulrike Felt (rapporteur), Brian Wynne (chairman), Michel Callon, Maria Eduarda Gonçalves, Sheila Jasanoff, Maria Jepsen, Pierre-Benoît Joly, Zdenek Konopasek, Stefan May, Claudia Neubauer, Arie Rip, Karen Siune, Andy Stirling, Mariachiara Tallacchini, on line (http://ec.europa.eu/research/science-society/document_library/pdf_06/european-knowledge-society_en.pdf). The report has also been pubilshed in Italian. Several materials about its public presentation are available on line www.fondazionebassetti.org/it/eventi/2008/02/science_and_governance_il_rapp.html

[23] For an in-depth study of the different national framework through which the problem of experimentaion at the frontiers of the life science is conveyed in juridical and philosophical language, see Jasanoff (2005).

[24] The closest example of a political experimentation about democratic choice and technoscience in agriculture is the project *Public Participation and Governance of Innovation*, led by the Bassetti Foundation and the Region Lombardy in partnership with Observa - Science in Society - in 2003-2004. The project involved citizens, entrepreneurs, consumers' associations and environmentalist associations in a public consultation about GM crops open field experiments. See Giuseppe Pellegrini, *Public Participation and the Governance of Innovation: main Results of the research project promoted by Lombardia Region, Irer, Bassetti Foundation and Observa,* http://www.fondazionebassetti.org/en/pages/2008/07/public_participation_and_the_g.html.

Bibliography

Adam B. 1998 *Timescapes of Modernity: Environment and Invisible Hazards.* London: Routledge

Bassetti, P. 2006 "L'innovazione responsabile, in AA.VV. Innovazione, creatività, responsabilità. Formare gli imprenditori del futuro". "*Quaderni della Fondazione Giannino Bassetti*". vol. 1: [page numbers].

Bucchi, B. 2006 *Scegliere il mondo che vogliamo*. Bologna. Il Mulino..

Bucchi M. and Neresini F. 2006 "Gli Italiani sono contrari alle biotecnologie? Alcuni stereotipi su opinione pubblica e biotecnologie in Italia", in M. Bucchi and F. Neresini *Cellule e Cittadini. Biotecnologie nello spazio pubblico*. Milano: Sironi Editore.

Dickens, P. 1996 *Reconstructing Nature. Alienation, Emancipation and the Division of Labour.* London: Routledge.

Drayton, R. 2000 *Nature's Government: Science, Imperial Britain, and the "Improvement" of the World*. New Haven: Yale University Press.

Escobar, A. 1999 "Steps to an Antiessentialist Political Ecology". *Current* Anthropology. 40 (1): 1-30.

Eugster, H. 1997 "Storia della razza bruna svizzera e della sua organizzazione". In Casanova, Eugster, Grüter, Harder, Herzog, Zogg, *100 anni. Federazione svizzera d'allevamento della razza bruna*. Zugo: Kündig Druck SA.

Franklin, S. 2007 *Dolly Mixtures. The Remaking of Genealogy*. John Hope Franklin Center. Durham Duke University Press

Fussell, G.E. 1984. *Landscape Painting and the Agricultural Revolution*. London: Pindar Press.

Galison, P. 2000 *The Ultimate Weapon: The H-Bomb Dilemma*. The History Channel.

Galli, C. 2003 "Editoriale. L'Opinione pubblica tra scienza e pregiudizio", *Bioetica*. 11(3): 421-426.

Grasseni, C. 2009 *Developing Skill, Developing Vision. Practices of Locality at the Foot of the Alps*. Oxford: Berghahn.

—. 2007 "Managing Cows: An Ethnography of Breeding Practices and Reproductive Technology in Contemporary Dairy Farming". *Studies in the History and Philosophy of Biological and Biomedical Sciences*. 38 (2): 488-510.

—. 2006 "Slow Food, Fast Genes: Timescapes of Authenticity and Innovation in the Anthropology of Food". *Cambridge Anthropology*. 25 (2): 79-94.

—. in press. "Reinventing Food: Beyond Consumption?" in J. Carrier (ed.) *Ethical Consumption: Social Value and Economic Practice*.

Hagendijk, R., and E. Kallerud. 2003 *Changing Conceptions and Practices of Governance in Science and Technology in Europe: A Framework for Analysis*. STAGE (Science, Technology, and Governance in Europe) Discussion Paper no. 2. Brussels: European Commission.

Hobart, M. (ed.) 1993 *An Anthropological Critique of Development. The Growth of Ignorance*. London: Routledge.

Jasanoff, S. 2005. *Designs on Nature: Science and Democracy in Europe and the United States*. Princeton: Princeton University Press.

Jones, S., 2005 "Transhumance Re-examined". *The Journal of the Royal Anthropological Institute*. 11 (2): 357-359.

Latour, B. 1988 *The Pasteurization of France*. Cambridge, MA: Harvard University Press.

—. 2004 *Politics of Nature: How to Bring the Sciences Into Democracy*. Cambridge, MA.: Harvard University Press.

Latour, B. and S.Woolgar 1979 *Laboratory Life. The Social Construction of Scientific Facts*. London: Sage.

Miller, D.P. and P.H. Reill (eds.) 1996 *Visions of Empire: Voyages, Botany, and Representations of Nature*. Cambridge: Cambridge University Press.

Neresini, F. 2000 "And Man Descended from the Sheep: the Public Debate on Cloning in the Italian Press". *Public Understanding of Science*. 9: 1-24

Orland, B. 2003 "Turbo-Cows. About the Production of a Competitive Animal in 19th and early 20th century". In S. Schrepfer and P. Scranton Piscataway (eds.), *Industrializing Organisms*. NJ: Rutgers University Press. Pp. 167-90.

Pellegrini, G. (ed.) 2008 *Technoscientific Innovation. Responsibility and New Models of Democracy in Science and Society Relationship*. Fondazione Giannino Bassetti/ Rubbettino.

Shapin, S. 2003 "Cheese and Late Modernity. Review of *Camembert, A National Myth* by Pierre Boisard". *London Review of Books*.11: 11-15.

Schaffer, S. And S. Shapin 1985 *Leviathan and the Air Pump: Hobbes, Boyle, and the Experimental Life*. Princeton: Princeton University Press.

Sibilla, P. 1991 "Uomini e animali in Valle d'Aosta e altrove". *Ricerca Folklorica*. 43: 91-102.

Tallacchini, M. 2002 "Cloni della scienza, cloni del diritto: il toro Galileo e l'ordinanza 5 marzo 1997", *Bioetica*, 10 (1):105-113.

CHAPTER SEVEN

HEALTHY MEAT AND NATURAL ANIMALS: NATURE, CULTURE AND DECOLONIZING KNOWLEDGE IN BOLIVIA

MAGGIE BOLTON

Introduction

This chapter resulted from a twelve-month period of fieldwork in the high Andes of Bolivia between 2003 and 2004. This took place in the community I call Rosario, in Sud Lípez province of Potosí department, where I have conducted ethnographic research over a number of years. The chapter was inspired by some conversations I had with llama herders about attempts by veterinarians and agronomists from development NGOs to persuade them to administer veterinary pharmaceuticals to their animals by injection in an attempt to control animal parasites. Development experts insisted on the need to administer pharmaceuticals to domestic animals to improve animal health and to prevent the transmission of parasites to humans through the consumption of meat. As in Nimmo's contribution to the volume, an important theme here concerns attempts by those versed in science to make the bodies of animals safe for humans to consume. Herders commented on the technologies using a variety of narratives about the detrimental effects the injections had on their animals: how "vaccines"[1] had "ruined" their llamas.

I examine here the role of animals at an intersection of science with indigenous knowledge in the context of the racial politics of the Andean region and government policy in contemporary Bolivia. One argument I make is that development professionals and their projects inadvertently collude with the urban *mestizo*[2] classes in what now seems a rather dated political project to assimilate indigenous people through modernisation to a homogeneous national ideal. I follow Law (1986:32) and other science studies scholars in seeing science as a potent arena for the pursuit of politics by another means, and power as an effect of diverse strategies to persuade others to accept a particular version of the natural world and impose "obligatory points of passage"[3] upon them. Science is translated from laboratory practice to impositions on society at large. In the Andes, it also passes through intermediaries in the form of development professionals who are brokers between science and indigenous knowledge and between rural and urban worlds. Such professionals are from the urban sectors of Bolivian society and at the same time are aware of the prejudices of those sectors towards rural people. While their aim is to improve the lives of rural dwellers, this improvement principally involves furthering their economic interests. They fail to acknowledge that rural people's motivation for livestock husbandry may depart from the economically rational. I find insights from science studies and in particular from actor-network theory to be useful in illuminating the entanglement of what are conventionally held to be science and the social and in showing how agency and knowledge production are distributed across collectivities even in a non-Western context, such as rural Bolivia. Nevertheless, I find that a focus on the motivations of llama herders ultimately highlights how Western notions of "interests" and the economically rational continue to underpin this theoretical approach.

Interculturality, Politics and Development in Bolivia

An approach to assimilate the *indian* population to a homogeneous national ideal through modernisation and education (and thereby to eliminate the supposed deficiencies of indigenous culture) was prevalent in the Andean countries in the early and middle years of the twentieth century.[4] In the 1990s, government policy towards indigenous people started to change in Bolivia, with constitutional amendments to demonstrate compliance with Convention 169 of the International Labour Organisation, on the rights of indigenous people, which formed part of a wider package of neoliberal reform. These amendments defined Bolivia as a multi-ethnic and pluricultural republic. Further constitutional changes towards an inclusive society, albeit stemming from a different political philosophy, have followed in the first decade of the twenty-first century under the left-leaning administrations of president Evo Morales.

The term *interculturalidad* (interculturality or interculturalism) started to appear in the discourse of intellectuals, government and non-governmental organisations at about the time of the first round of constitutional changes. While it is a slippery term that means different things to different users, some general observations concerning its use can be made. It originated in the language policies of Latin American countries towards indigenous languages in education, and was coined to create a distinction from the approach, labelled "bicultural", promoted by the Summer Institute of Linguistics (an evangelical Christian Organisation) (Hornberger 2000).[5] As the term spread to politics and other types of education, *interculturalidad* also became used to differentiate a Latin American model of plural societies from the *multiculturalism* of Western nations. In the West, multiculturalism aims to respect, guarantee the rights and offset the disadvantages of minority groups. In Bolivia, under the growing influence of indigenous identity politics, *interculturalidad* became associated with the ideas of "decolonizing the nation", and of collaborations and dialogues between its different populations which were to lead to the formation of new cultural identities (PIEB, n.d.). While the Morales administration refers less to *interculturalidad* than did previous regimes, the notion of "decolonisation" has become central to its philosophy.

It is debatable how successful *interculturalidad* and decolonisation have been in Bolivia in bringing indigenous and *mestizo* populations towards collaboration and dialogue. Hornberger (2000) notes that Bolivian language policy is more pluralistic than are such policies in other Andean countries. However, ambitious attempts to decolonize democratic politics under the first Morales administration have been diluted in the face of right-wing opposition.[6]

A decolonizing or inter-epistemological approach would seem appropriate for agricultural development involving the activities of indigenous people. It is an arena in which dialogue could reach beyond the conventional bounds of what is considered cultural and seek accommodations between indigenous knowledge and science. Nevertheless, my observations of a livestock development project in the Sud Lípez highlands were that developers educated in scientific disciplines made little attempt to establish any real conversation with Andean herders about their different understandings and expectations of animals, improvement, alimentation, medicines or even the long-term goals of livestock husbandry. They were more concerned to correct deficiencies they perceived in Andean animal management and to persuade herders to adopt "scientific" methods of raising animals, as well as to counter the prejudices of potential consumers of llama products living in urban areas. In livestock development, it is as though the debates surrounding cultural accommodation and decolonisation had never taken place.

Some commentators have drawn attention to the colonial connotations of development[7] and to contradictions between science and indigenous understandings in the Andes.[8] Nevertheless, science and development do not seem to have been subject to the sort of decolonizing approach that is increasingly considered necessary in other areas of life. Interculturality is deemed appropriate to language policy, politics, and certain textual practices, to which anthropologists have drawn attention (Arnold and Yapita 2000), but not to science. This is because science is conventionally held to lie outside culture and involve action on nature. In assuming that different cultures should collaborate and enter dialogue, I suggest that most promoters of interculturality implicitly assume that these cultures are different to, but congruent with, the Western model, and that "nature" is a constant existing outside and as a binary opposite to each culture: how Bruno Latour (1993: 103-109) describes conventional cultural relativism. Policy makers and development professionals have yet to take what Latour would term a "symmetrical" anthropological approach relativising nature and culture–or whatever other categories are appropriate–across different "collectives" or cosmologies. They have yet to take a critical stance towards Western science as a particularly universalising vision of relations between humans and non-humans formulated by situated humans working within a specific intellectual tradition (Latour 1993: 107), although such an approach would be in tune with current government policies of decolonisation.

In what follows, I want to try to take up Latour's call for a symmetrical anthropology. I explore Andean people's understanding of animals, their bodies and the medicines injected into them in an attempt to understand why herders should distrust the medicines promoted by developers. I start by exploring Andean ideas about different kinds of animals, purity, cleanliness and what is understood to be "natural". I then highlight the tensions between rural and urban worlds, and particularly examine urban ideas and regimes of hygiene as these regimes impact upon rural people and llama meat. In the light of these discussions I examine the actions of the developers and the methods they advocate, my goal being to show that these actions cannot be considered simply as action on nature, but are themselves caught up in a mesh of racial politics, international restrictions and regulations and ideas of what it is to be modern. Towards the end of the paper I return to the narratives of the herders and their experiences of the use of vaccines on their animals. Through thinking about the vaccines in relation to their goals and motivations, I conclude by asking whether science studies itself requires "decolonisation".

Animals, Animation and Meat in the Sud Lípez Highlands

Herders in the harsh, arid highlands of Sud Lípez keep llamas but not alpacas, which prefer moister grazing. Ownership of llamas is hugely important to them. They are a symbolic marker of identity[9] and both a symbolic and material marker of individual wealth that can withstand the human-made catastrophes of economic recession and political turmoil to which Bolivia is particularly prone. Llamas provide people with fibre for weaving–considered by many to be the original motivation for animal domestication in the Andean region (Dransart 2002: 11-12)–although lipeños (people from Lípez) now weave much less now than they used to. Llamas used to serve as beasts of burden over long distances, but have now been all but superseded by motorised transport. They also provide people with meat and hides. A relative increase in the importance of llamas as producers of meat has come about with the decline of their importance as providers of fibre and transport.

Lipeños eat small quantities of meat on a daily basis, usually in soups, and they eat and share very large quantities of it on festive occasions. As well as llama, they eat sheep and goats, a few wild mammals and birds and, when they hope nobody is looking, donkeys. They sell some llama meat, both fresh and dried, to neighbours and to retailers in the cities of Uyuni and Tupiza, although they are reluctant to sell live animals because this is a separation of parts from a unit (the herd) that they believe should retain its conceptual integrity.[10]

Andean cosmology does not fit well with European ideas of culture as a purely human realm and nature as cultureless and amenable to domination by humans. Landscape features are held to be animate and powerful supernatural beings, to whom humans must show respect, and who have lives and social relations similar to those of people. In a well-known article, Olivia Harris (1980: 85) suggests that rather than distinguish between nature and culture, for Andeans a more salient distinction is between the domestic, cultivated spaces of humans and the wild spaces of the mountain spirits. In this volume, Dransart suggests that a parallelism exists between llamas and people: llamas and humans have different characteristics, but both are attributed autonomy, engage in social relations appropriate for their kind and have their own lineages. Taking this sort of argument a step further, Abercrombie (1998: 184) has suggested that, for Andeans, llamas are "suitable subjects to think about in attempt to change their own social relations" and Olivia Harris (n.d.) that human groups in the Andes have taken llama herds as a model or template for their own social organisation. Humans care for llamas on a daily basis, and llamas reciprocate when they stand in for humans in sacrifice and so mediate between humans and the supernatural landscape powers. We might say that while Andeans hold animals and humans to be different, they extend features that we hold as specific to human culture to both animals and the landscape.

Natural and Unnatural Animals

While the Western notion of nature may not be appropriate to Andean cosmology, lipeños do use the Spanish word *natural*[11] to describe both animals and people: an animal or person that is *natural*, or *original*, is native to a particular region. They make a distinction between animals and people that are *naturales* or *originales* and those that are not.[12] Indigenous highland people are *originarios,* having a close connection to the land on which they live, while city dwelling "whites" and *mestizos* with no affiliation to the land are *q'aras* (literally "naked"). Lipeños consider their llamas to be more *natural* to Lípez than they are themselves. Despite the fact that they have recently started promoting themselves as *originales* for political reasons, many people recognise that their own ancestors came from elsewhere to work the region's silver mines in the colonial era. Llamas, by contrast, are considered to have originated locally at a place called Llamayoqpata ("on top of the llama owner") where a *samiri*[13] (generative power) caused them to emerge from the earth. My landlady, doña Natalia, told me that in the mythical past, when humans and animals were more alike, llamas spoke Aymara, as did lipeños back in the sixteenth century, before mine labourers introduced Quechua to the region.

As well as considering them *naturales*, lipeños regard llamas as the cleanest of animals. Llamas are clean because they deposit their droppings together in heaps, are particular about what they eat, and pick up their feet carefully as they walk. As a consequence of such exemplary habits, they do not smell unpleasant when they are butchered. A llama slaughter is a social and ritual occasion. Following slaughter, pieces of a llama's ears, toe clippings, heart and liver are interred as an offering to the earth in a hole in the ground as sacrificial tribute. Such tribute helps sustain, or perhaps even constitute, a link between humans and the land they inhabit.

Sheep are owned by lipeños in smaller quantities, and are not considered as *natural* as llamas: they are held to have spoken Spanish in former times, reflecting their European origin. However they are also credited in myth with a local place of origin a lake on the Argentine side of the nearby international border. This ambiguity reflects their status as foreign imports, yet wool-bearing animals that can be assimilated readily into

Figure 7-1: Slaughtering a llama at a Sud Lípez homestead (photograph: M. Bolton)

an Andean way of life. They are not considered to be as clean as llamas. While lipeños have no qualms about eating them, I heard one of doña Natalia's neighbours comment during slaughter that they smell of onions, and adding that they are *cochinos* (pigs) that eat indiscriminately.

Apart from llamas, sheep, wild vicuñas and a few goats, the lipeño countryside is grazed by numerous donkeys. These animals have been left over from the era before widespread motorised transport, when herders used them as pack animals: although llamas are considered best for long distances, donkeys are sometimes preferred for shorter trips, as they can carry heavier loads. Their owners round them up once a year to have them counted by community officials and they pay a local tax on them. These days, donkeys rarely work and wander the hillsides largely untroubled by humans. Lipeños express much more ambivalence about donkeys than they do about either llamas or sheep. They are not native, and while both llamas and sheep are considered to once have been "more like" humans, doña Natalia stated directly that donkeys used to be people.[14] Their classificatory troublesomeness stems from their teeth, which transgress a boundary between what is considered a characteristic of humans and that observed in other domestic animals: donkeys have upper and lower dentition, like people, while llamas and sheep have a hard palate in place of an upper row of teeth.

While llama meat is considered clean and good to eat, and parts of llamas enter into ritual observances, people express revulsion at the thought of eating donkey: the idea disgusts them, they say, and makes them feel sick. They also comment that the meat stinks. One woman told me that the consumption of donkey meat was expressly forbidden by the Bible. Nevertheless, many will tell you in confidence that their neighbours eat donkey, or sell their meat to unsuspecting customers as llama.

In fact quite a lot of donkey meat is consumed. The less well off can feed a family cheaply by buying and slaughtering an old animal. Those who still own large numbers of donkeys eat them surreptitiously. I was told that donkey was once served covertly in Rosario to a meeting of the peasants' union,[15] and there is quite a trade in the dried, salted meat (*ch'arki*). This trade started with commerce across the Chilean border–where, lipeños claim, the consumption of donkey *ch'arki* is accepted and normal–and has now spread to Bolivian cities like Tupiza and Uyuni, where it is sold covertly as llama. While the slaughter of a llama is always a social and ritual occasion, donkeys are slaughtered without ritual observance and unused parts of their carcasses are left around for dogs to eat.

Dirty Llama Meat in the City

Studies in anthropology and cultural history abound with examinations of the symbolism of dirt, hygiene and their relation to systems of classification (e.g. Douglas 1966) and of the use of hygienic measures as a means of social control (e.g. de la Cadena 2000). Latour (1988) in his study of French hygienists, *The Pasteurisation of France,* is nevertheless impatient with accounts that explain hygienic concerns and measures in terms of factors that are purely social (or cultural). In what follows, in a similar manner to Whitehouse in this volume, I hope to show how science is entangled with politics, or rather how hygiene, vaccines and parasites become enmeshed in the racial tensions of the Andes.

An examination of hygienic concerns and meat in urban areas helps illuminate some of the logic of former policies for the assimilation of indigenous people. For herders, llamas are the cleanest of animals, but in cities their meat has long attracted attention for being just the opposite. Meat sold in Bolivian cities has a hierarchy: central markets have stalls selling beef, pork and chicken, while lamb and mutton are sold at more out-of-the-way spots. Until recently the sale of llama meat was expressly forbidden by municipal authorities in most urban markets on grounds of hygiene and most meat that found its way into cities was sold covertly.[16] During the Third Worldwide Camelid Congress, held in Potosí in 2003 an event that achieved high status in the city due to the number of foreign academics in attendance, the city mayor commented to a local newspaper about his own rural childhood, when: "It was little short of a crime to sell llama meat in the city, we had to do it clandestinely because if we were caught it would be confiscated".[17]

It is here that we should perhaps move from Bruno Latour's call for us to relativise nature and culture–which he considers the Modern Constitution to have separated in an activity he terms "purification"–to Mary Douglas's earlier work on "purity" and classification.[18] That meat should become the object of a discourse of hygiene and subject to regulations is perhaps unsurprising. Famously, Douglas examined how, for ancient Hebrews, certain animals that fit uneasily into their schemes of classification were considered dangerously transgressive of boundaries and prohibited as food (Douglas 1966: 41-67). Only those considered "clean" were suitable to transgress that most radical boundary between the inside and the outside of the human body. As we have already noted, lipeños themselves are concerned about boundaries and the cleanliness of the animals they eat.

So, why should the animals that exemplify cleanliness to lipeños be so reviled by city dwellers? After all, an idealised representation of a llama appears on Bolivia's national coat of arms. The problem seems to be that for urban people, real llamas (and their meat) when they enter the city transgress a significant boundary: between rural communities where animals are raised and urban markets where meat is sold. This remains a boundary between rural "indians" and urban "whites" and *mestizos*. Llamas are the animals most closely associated with indigenous people and the incursion of their meat into cities represents an intrusion of the rural, indigenous world into urban, "white" and *mestizo* space.

In an article on the interplay of biomedicine, hygiene and ethnicity in early twentieth-century Bolivia, historian Ann Zulawski has shown how the Bolivia's urban population associated rural indigenous people with lack of hygiene and propensity to diseases that at times threatened to engulf the cities (Zulawski 2000: 116). City-dwellers, particularly the physicians of the period, laid the blame for this perceived lack of hygiene squarely on the "culture" of rural Andean people rather than on biological substance, which they to some extent shared with them (Zulawski 2000: 118-119). For the more socialist among them, the solution lay in eliminating the contaminating culture and absorbing indigenous people into the national working class, a sentiment that persisted after the National Revolution of 1952.

As discussed earlier, assimilation as a general policy has been superseded by interculturality and corresponding changes have occurred in the urban attitude to llama meat. As development anthropologist Kevin Healy (2004: 32) notes, by the late 1990s most municipal legislation discriminating against the meat had been repealed. Llama meat is now available in cities and has made some converts, but an aura of dirt still attaches to it for many urbanites–as was apparent on the face of a friend in Sucre when I presented her with a pound of llama *ch'arki* that I had received as a parting gift from a lipeño herder, and that HM Customs forbade me from carrying back to the UK

Present-day urban discourse surrounding llama meat illustrates the persistence of the association of dirt with indigenous lifestyles and the barriers that exist to real dialogue between rural and urban populations. Llama meat, coming from animals so closely associated with rural Andeans, seems to exemplify for urbanites all that they deem unhygienic or even un-human about the way of life of rural dwellers. As Olivia Harris (n.d.) has noted, a widespread term of racial abuse in Bolivia today is precisely "llama", an insult that links the endogamy of rural groups with animal behaviour lacking any incest prohibition and Healy (2001: 192) reinforces the animal association when he comments that llama meat is considered by city-dwellers to be "Indian food, more suitable for dogs than humans".

Urban concern about the dangers of llama meat is rationalised in two separate, but related discourses. Firstly, as I alluded above, it is claimed that meat is produced in unhygienic conditions: city dwellers' sensibilities are offended by the idea of llamas slaughtered in corrals full of dung. Secondly, it is believed that the meat carries disease.

Rumours abound about the dangers of the meat, including, in another reference linking the animals to unregulated sexual behaviour, that it can give you syphilis.[19] The usual justification given by city dwellers for avoiding it, however, is the presence in some meat of small, white cysts, which they claim are *trichina* (*Trichinella spiralis*) the harmful nematode parasite found in pork. This was the concern voiced by my friend in Sucre. The cysts in llama meat are not *trichina*, but it serves urban sensibilities to believe they are and, as Healy (2004: 31) notes, fear of contagion from llama meat, the exemplary *indian* food, is an outward expression of racism on the part of the urban population towards indigenous herders. Their discomfort may be exacerbated by the presence of a parasite that can transgress the boundary between animals and human beings, undoing the work of purification in separating the modernist concepts of nature and culture. It is almost certainly made worse by an organism that fails to distinguish between themselves and indigenous people, whose culture they all but relegate to the realm of animal behaviour and nature.

Making Llama Meat Safe for City Dwellers

Between 2003 and 2006 an international NGO financed by the Italian government and the European Union implemented a livestock development project with llama herders in the southwest of Potosí department, the region in which Sud Lípez lies. The aim of the project was to transform subsistence herding into a commercially profitable operation for herders, and encourage herders to produce fibre and meat for national and international markets. This was to be accomplished by technical personnel with scientific backgrounds, mainly from Bolivia's cities, working with local facilitators whose training had been financed by previous livestock-oriented projects. Together they would persuade herders of the need to commercialise their activities and adopt a model of animal husbandry based on scientific principles, involving animal improvement, genetic management and the use of veterinary medicines, similar to that used by the intensive livestock farmers of industrialised countries. Insofar as llama meat was concerned, in the short term, developers hoped to build on the existing trade in which herders engage and to make llama meat a food of choice for urban Bolivians. In the longer term, they look towards an international market, and identify the Slow Food Movement,[20] with its interest in traditional foods and support for artisan producers, as an organisation through which llama-meat products such as *ch'arki* could be promoted.

Here my concern is with the animal-health and hygiene measures that developers promote and attempt to make into obligatory points of passage for herders as the developers attempt to impose their version of the natural world on them. I also explore how such measures are enmeshed not only with the bodies of animals, microorganisms and other parasites, but with legal regulations, prejudices and political relations between the different sectors of Bolivia's population.

The control of foot-and-mouth disease (FMD) provides an example of this entanglement. Over the course of my fieldwork, vets from the government agency SENASAG (*Servicio Nacional Agropecuaria e Inocuidad Alimentaria*) were gathering data with the aim of declaring the Western side of the country a foot-and-mouth free zone in line with the requirements of the international regulatory body for animal health, the

OIE[21]. Their zeal in pursuing such certification puzzled me, partly as a result of herders' own suspicions both of SENASAG's practices of enumeration (see Bolton 2007) and of animal vaccination.[22] Such zealousness was further puzzling because papers published in veterinary science that mention foot-and-mouth in camelids seemed to agree that llamas have low susceptibility to the disease, when infected display only mild symptoms and pose practically no risk of transmitting the virus to other livestock (see for example, http://www.llamas.co.uk/Pages/fmd.htm).

As Law (2004) has noted, achieving foot-and-mouth free classification is advantageous for potential exporters, but to aspire to this status and introduce the controls necessary to satisfy the OIE is enormously costly.[23] It involves setting up disciplinary regimes in the form of systems for disease surveillance, reporting, control and eradication–all expensive procedures that far exceed what could be justified by the export of a small quantity of llama *ch'arki* from a country of limited resources like Bolivia. However, the drive for foot-and-mouth free certification in llama-herding areas may be a step toward achieving a wider goal of FMD-free status for the entire national territory, which could be advantageous not so much to indigenous herders but to the cattle ranchers of the eastern lowlands, hoping to gain entrance to the international beef market supplying hamburger chains. Notably, it also provides employment for a number of technically-qualified people and increases the visibility of the organization that employs them.

Attempts to control one animal-borne disease thus did not involve only the animals affected by and the virus held responsible for an illness. They also involved international regulations, regimes of control, relations between the developed world and the global South, global commerce, fast food, a Bolivian government agency, the economic interests of the veterinary professionals it employs, and the (extremely strained) relations between the non-indigenous eastern departments of Bolivia and the predominantly indigenous highlands.

The NGO responsible for the livestock project supported SENASAG's efforts on FMD and also promoted a number of health and hygiene measures concerning animals, their diseases and the problems faced by llama meat on the domestic market. In the 1990s, city-dwellers' revulsion at the thought of eating llama meat produced in unhygienic conditions had motivated the much-publicised construction of the first abattoir designed for camelids in the town of Turco in Oruro department (Healy 2001: 211-216). Although such an installation far exceeds the budget of Sud Lípez municipalities, NGO experts working there recommended the construction of low-cost platforms (*playas de faeneo*) to raise slaughter above the contaminating rural ground.

In tandem with these changes, an altered perception of llama meat is being accomplished by skilful publicity, packaging and labelling. A number of studies have drawn attention to the healthy properties of the meat–low fat and cholesterol content–and in the handful of instances where NGO facilitators have persuaded members of rural communities to start small businesses (*microempresas*) producing llama *ch'arki*, they have encouraged them to package their product in cellophane envelopes with lists of contents on their labels, like supermarket packaging. Such packaging not only separates llama meat from any suggestion of indigenous culture, but also suggests "technification", hygienic production and scientific analysis, even if the tables of contents are copied from rival businesses and NGO reports.

Moving from hygiene in the production of meat to the control of animal parasites, NGOs working in the area in the 1990s constructed anti-parasitic baths for llamas and promoted animal-dipping regimes to combat *sarna* (sarcoptic mange) caused by scabies mites. This met with general approval from herders who describe mange as a disease resistant to their traditional remedies. One woman told me that the condition was a punishment from the ancestors (*almas*) and that the only remedy was to kill the affected animal and place its bones in the river.[24] However, other people observed that an increase in the incidence of the condition had accompanied the recent proliferation of wild vicuñas, a formerly endangered species now enjoying legal protection. [25] The very need for new regimes of hygiene is hence entangled with measures aimed at wildlife conservation which, as Whitehouse has observed in this volume, can have unintended and unpredictable consequences. Environmental concerns have also entered the fray as a factor leading to the promotion of antiparasitic injections as a means to control mange, since developers have started to worry about the damaging effects of the insecticidal chemicals used in dips on the surrounding countryside.

The cysts in meat that repelled city-dwellers are an obvious target for action. As stated previously, these cysts are not *trichina*. They are a parasitic protozoan species *Sarcocystis aucheniae*. Infection with this organism seems to be largely asymptomatic both in llamas and people, although some experts claim that people unaccustomed to such fare can feel nauseous after eating infected meat. Most herders consider cysts in meat to be "normal". They are known locally as *arrosillo*, since they resemble grains of rice (Sp. *arroz*), are found most frequently in the meat of older animals, and, lipeños assert, no one eating them suffers ill effects.

A few days after we had eaten a meal together of meat speckled with the white, granular cysts, I asked one of my *compadres*[26] whether they might be harmful. He commented:

> "No, no... How is it that we have lived here for hundreds, thousands of years? How come we don't die? Of course I've eaten this, all the time! Even the other day, we ate this meat together! Damn it! Nothing happens! Nothing happens!"
> [Carlos age approx 45]

Sarcocystis aucheniae has a complicated two-stage life cycle that involves a phase of asexual reproduction in the muscles of camelids, its intermediate hosts, and a phase of sexual reproduction in the gut of a carnivorous definitive host. The cysts egested by the carnivorous host in faeces are ingested by grazing llamas and the carnivorous host becomes re-infected through consuming llama meat or blood. It is not an easy parasite to tackle. In the most widely read text on camelid medicine outside the Andean region, Murray Fowler, a veterinary expert based at the University of California, states that there is no treatment for the condition, and that it can be addressed only by preventative management (Fowler 1999:212-3).

Preventative management was advocated by both NGOs involved in camelid development and SENASAG. A poster campaign prominent at llama shows during my fieldwork, identified dogs, as the definitive hosts, and urged herders to keep them away from llama meat and blood or to destroy them. Despite being ostensibly aimed at herders, this campaign seems to have been largely for the consumption of the urban population–who thus witnessed SENASAG and the NGOs making efforts to make meat safe–since only a small number of herders attend shows in urban centres, while urban people attended events like the prestigious World Camelid Congress (see Bolton 2006) in droves. Developers did not push preventative regimes involving the elimination of dogs very actively with herders. They knew that most herders like to keep a dog with their herd to scare away foxes (serious predators of newborn llamas) and often feed the dog a bowl of blood after an animal is slaughtered.

That the poster campaign responded to urban sensibilities perhaps explains its failure to mention the possibility that humans might also act as definitive hosts for the parasite: at least some *Sarcocystis* species are known to be zoonotic.[27] In the countryside, where latrines and sewers are absent, the parasite could well be spread by people who ingest cysts, which can survive in undercooked meat, and who defecate where llamas graze. I did not appreciate all the details of the life cycle of the protozoan until after my return from the field, and so did not question the authors of the poster campaign about the possibility of human transmission, but would suggest that to mention it in a campaign likely to be viewed by city dwellers would confirm urban prejudices about the animal-like behaviour of indigenous people and would undermine attempts to promote llama meat as produced under conditions of hygienic control and disease management.

"The Vaccines Have Ruined Our llamas"

A large part of the effort to control animal-borne parasites has taken the form of injections of broad spectrum anti-parasitic medications, such as ivermectin.[28] Injections combining just such an antiparasitic with a fortifying tonic formed part of a highly publicised package of aid supplied to communities in the south of Potosí department by international aid organisations following a disastrously heavy snowfall in 2002, which was the first occasion upon which most lipeños injected their llamas with pharmaceuticals. The snow had a devastating effect on many people's herds, as many animals died in the blizzard and still more perished from starvation in the months that followed as pasture became scarce. As well as saving the lives of animals weakened by hunger, the pharmaceutical injection drive aimed to encourage herders in the systematic use of antiparasitic medication.

The pharmaceuticals were described to herders by NGO facilitators as vaccines (*vacunas*) to bring them into the realm of the familiar for rural people accustomed to taking their children to the health post for vaccinations. However, "wormers" would be a more accurate description. They were also promoted vaguely as being effective against both internal and external parasites. Such promotion amounts to what Law (1986: 14) would term a distorting simplification: like herders, I gained the impression that this included the cysts found in meat. I was surprised on returning from the field to discover that ivermectin is not effective against sarcocysts: it attacks the nervous systems and so is effective against roundworms, tapeworms and skin-borne parasites like mites, in which a nervous system is present, but not against single-celled protozoans.

Lipeño herders do acknowledge that their llamas suffer from disease. People I knew became visibly upset if one of their animals fell sick and died, and older women, in particular, use local plants to treat a range of animal symptoms from diarrhoea and fever to failure to urinate. They also recognise some internal parasites as problematic and talk about their animals having *gusanos* (maggots or worms), some types of which have specific Quechua names. In former times pack animals travelling to the valleys on journeys of exchange were considered particularly susceptible to *gusanos* and, in the course of rituals held on return to the highlands, would be given a drink of corn beer laced with the highland moss, *yareta*, to eliminate them. Effective local solutions to the problem of intestinal worms hence existed prior to the work of NGOs.[29] However, rather than encouraging their use on a more systematic basis, developers saw a need to introduce herders to the use of pharmaceuticals, and thus attempt to bring herding practices into line with animal husbandry in the developed world. Furthermore, this was done under the guise of offering aid to herds weakened by a major crisis.

Although some herders accept the need for the "vaccinations" and consider their use to indicative of their own modern attitudes, others have reacted to them with suspicion. Some have become increasingly articulate in expressing opposition, often in the light of their experience with the free handouts they received in the wake of the snowfall.

In 2002, most herders allowed their animals to be injected partly because the medicines were freely available, but also because they believed they would restore fat to emaciated llamas. The first complaint people had against the injections given was rooted in their initial high expectations. Some people considered the injections to be "vaccination against thinness",[30] and were severely disappointed when they failed to prevent their malnourished animals from dying. When animals continued to die, they suggested that the medication must have been "too simple". Even the local NGO facilitator, trained as a veterinary technician, suggested they might have been watered down (*adulterados*).

A very distressing observation for the herders was that not only had adult animals died from starvation, but also that there were increased incidences of stillbirths and birth deformities among their llamas in the months that followed. Herders identified the medicines as a significant factor behind these phenomena.[31] A common refrain that came to be heard was that "…the vaccines have ruined our llamas". The Spanish verb *arruinar* used by herders translates into English not only as "ruined" or "destroyed", but also as "blighted", a term that conveys the idea of a general contamination affecting the animals. Individuals described the effects in different ways. For some it had lessened the nutritional value of the animals' meat for human consumers. For others it had compromised llamas' status as "natural" animals. For yet others, it seemed to have caused them to have the appearance of people. The following exchanges were typical of people's reactions:

I applied these medicines and they have turned out something bad. The young emerge sort of twisted, you could say, so that some cannot walk, or the animal has no milk. All these situations happen….
[Mauricio, age approx. 62]

In camelids, we give some injections, a few… some medicines, but not everyone accepts them. For example, my father and my own family… and I myself, personally, I don't want much [of this]. Science also has its negative side, doesn't it? Because our llamas and camelids are *originarios*, *naturales* and they don't know any type of injection… Thanks to science we get lots of vegetables, all improved and good to the eye, don't we? But they're poor inside.
[Geraldo, age approx. 27]

...in years gone by, we didn't even know [antiparasitic] baths, [the llamas] were just like this, because it is better that that the llama meat… that we don't do those things. It has more vitamins, and we say that we need only a little meat to go in the pot. Now that we do these things, we say that nothing happens with the baths, but regarding the internal [vaccines]… it seems to me that this is not obviously natural. And I say that the meat is nothing, it doesn't have any vitamins, it's more or less like that. We throw some in the pot and nothing… we put in lots of meat… and it's the same. But in years gone by, we used to put a little meat in the pot and it gave it flavour, but not now. This is a consequence of some of us managing [our animals] using medicines internally.
[Mauricio, age approx. 62]

Tomasa: We got them to give the llamas vaccines, didn't we? But last year, after we had injected them, the llamas were born differently.
Maggie: How were they born?
Tomasa: *They had faces almost like people*. They were born without a jawbone… For example, in our case…
Maggie: Without a jaw?
Tomasa: Yes, without a jawbone. That's how they were born, and our parents think that the vaccine has affected them, don't they? So, that's almost all of our parents who don't want to vaccinate. Of course, we [the younger generation] could vaccinate them, but they [our parents] are more the owners of the llamas, aren't they. He who cares more is the owner… so, that's what happened last year. In our llamas it happened too… We've seen it.
Maggie: Did you? What were the young like?
Tomasa: Without a jaw, they didn't have… How can I say? They couldn't walk, nothing, it's like their hands were tied, and they couldn't move…
[Tomasa, age approx. 38]

As the quotes from Tomasa, Geraldo and Mauricio indicate, many people thought the injections had damaged their animals. Even younger herders, like Tomasa, who considered the use of pharmaceuticals to be prestigious and modern, and her brother Geraldo, concurred that the "vaccines" were responsible for the harm they suffered. Their llamas had become "poor inside" like vegetables grown with unnatural fertilisers

Natural Animals, Humans and the "Fat-Snatcher"

An examination of lipeño preoccupations with internal animal vaccines, and why their use might be resisted, helps throw light on the more fundamental matter of herders' motivations for keeping animals and the goals of their animal husbandry. A good starting point for this is the circulation of fluids both within animal bodies and in wider circuits of exchange. Medical anthropologist Joseph Bastien (2002: 173) proposes that Andean people have an ethnophysiological preoccupation with the loss, intrusion and circulation of fluids in the human body and that this influences their acceptance or otherwise of certain medical procedures, such as blood tests that draw fluid from it. The intrusion of fluid caused by the administration of pharmaceuticals by injection to llamas may similarly upset a delicate balance. Lipeños regard the medicines as vaccines. Although they take their children to the medical post for vaccination, they may not be entirely reconciled to this: in an echo of the comments made about llamas and vaccines, I was told by one man that in former times, when people "did not know vaccines", they were much stronger and did not get sick like people do today. The intrusion of vaccines is understood to have already affected the balance of fluids within the bodies of humans.

As mentioned above, llamas are considered the most "natural" of domestic animals, as well as the cleanest and best to eat. Their cleanliness stems from their tidy disposal of the body's waste products and from being particular about what they allow into their bodies in the first place. As Geraldo pointed out, "…our llamas are *naturales, originales* and they don't know any type of injection". Injecting them with industrial pharmaceuticals compromises their natural and clean status, making them less valuable as food to humans, so that more and more meat is needed to make soup flavoursome. Although no one stated so directly, it seems that llamas treated with vaccines likewise may not nourish the powerful landscape deities adequately through sacrifice.

Sacrifices or payments to the landscape frequently take the form of llama fat: as *llumpaqa* (mixed with flour), fat is sprinkled on the earth in rituals. It forms part of the offerings burned on festive occasions. Miniature

Figure 7-2: Offering miniature llamas to the earth on the morning of the llama-marking ceremony (Photograph: M. Bolton)

llamas of fat and flour are modelled on the eve of the llama marking ceremony, when herders ask for the multiplication of their herds, and also after a death, when they carry a dead person's possessions to the next life. As Canessa (2000) notes, fat, not blood, is the vital substance, a source of agency and creativity, for Andean people. It enters into circuits of legitimate exchange which form the basis of human social reproduction: humans nurture landscape deities with llama fat and in return receive the products of the earth, which through consumption they transform into the fat of their own bodies.

Throughout the Andean region, people fear losing fat. One particular genre of myths concerns people, usually white men, who prey upon indigenous people and steal fat from their bodies. These fat-stealing ghouls, or *lik'ichiris*[32] as they are known in the south of Bolivia, leave their unwitting victims to die, as they divert fat from its legitimate circuits of exchange and commercialise it in global markets. Such stories illustrate the unequal relations of exchange that characterise dealings between indigenous people and outsiders. They have also changed with the times: while colonial *lik'ichiris* greased the moulds of bell-founders, these days they grease the machinery of international capitalism, diverting the fat into industries such as space technology… and the manufacture of pharmaceuticals (Gose 1986).

It was concern about fat that sparked the rumours about the medicines: they failed to restore fat to emaciated animals after the snowfall. Following injections, llama meat became less nutritious than before, a property that for lipeños is strongly associated with its fat content. On slaughtering a llama, someone inevitably comments on whether the carcass is fat, thin or "regular" and during my fieldwork they were inevitably described as thin. Contaminated llamas were unable to produce the fat necessary to sustain either humans or their relations with their environment. Some lipeños even laid the blame for the snowfall on the neglect of sacrifices, or payments of fat, to the most prominent mountain of the region, Cerro Bonete. Tomasa's observations, that llamas born after the injections had the faces of people, appear less arbitrary in the light of a widely held suspicion that the pharmaceutical fluids injected originate in an industry running on stolen, human fat.

The development project to commercialise llama meat may have been on precarious ground from the outset, aiming to convert the animal at the heart of sacrifice and reciprocal relations with the landscape into a commodity. It also relied upon the assumption that the interests of herders were those of rational economic individuals seeking commercial gain. It thus ignored some of the outcomes of llama husbandry desired by indigenous people but which did not fit with developers' conceptualisation of the activity, namely the maintenance of relations with a variety of supernatural beings and the achievement of a sort of balance between what they understand to be the different inhabitants of the Andean landscape. Herders value the maintenance and renewal of these relations as well as monetary profit, but relations between humans, mountains and animals are not taken into account by agricultural development, especially when there is a need to be seen by the outside world as having introduced modern, hygienic methods.

Conclusions

One of my concerns from the outset of this paper was to look at the workings of agricultural development and the science contained therein in the light of the current political concerns about decolonisation and the revalorisation of indigenous cultures and knowledge in Bolivia. An examination of measures taken against animal-borne diseases by NGOs working with highland llama herders has shown just how far away from collaboration, dialogue and decolonisation such projects still remain. It has also shown how enmeshed the supposedly scientific control of animal disease is with the interests of the developers themselves, racial politics and concerns about the environment and conservation, themselves a combination of what is conventionally held to be science with the political.

All the above factors leave little room for accommodation between science and indigenous knowledge. Herders are still expected to cast off their ignorance and subscribe to the views about animals, their diseases and the prevention of disease promoted by developers and there has been little attempt to build on herders' traditional knowledge even when this has direct, practical application according to scientific criteria–as is the case with traditional worming medicines. Insofar as developers seek to "sign up" herders to Western science, they are engaged in a project akin to the political projects of assimilation of yesteryear, and have yet to take on board more recent ideas about decolonizing knowledge. A decolonising approach would involve working with herders and seeking ways to integrate veterinary science with their existing knowledge in a similar manner to some recent initiatives in the field of human medicine (see for example Bastien 1992).

In this paper I have attempted to make use of some insights from science studies. Such a perspective takes a critical stance towards scientific knowledge as well as to other knowledge that is conventionally termed social. It helps show how developers make parasites–microbes and visible organisms such as protozoans–appear as actors and how the technical personnel of NGOs aim to become an obligatory point of passage in the fight against these enemies. It also, more generally, does a good job at showing how agency and creativity are distributed rather than located with individual human beings, bringing Western ideas about individual authorship under scrutiny and creating resonances with anthropological scholarship on thinking in indigenous societies in diverse parts of the world (e.g. Leach 2007). Nevertheless, at its core still lie assumptions about the "interests" (if not the conscious motivations) of the actors involved, whether human or non-human, for example, the scallops in Michel Callon's much acclaimed essay on St. Brieuc Bay (Callon 1986). The approach relies upon the idea of a network becoming stable as actors align their interests, and these interests all too often appear akin to those of a rational economic actor in the Western world, or a "stakeholder".

The case of llamas, herders and vaccines could be taken as a negative example of the above effect, as Callon takes his St. Brieuc Bay example, when actors fail to align their interests. However, I am reluctant to reduce herders' concerns about prioritizing relationships between humans, animals and animate-landscape beings to this category or to attribute specific interests to llamas or other entities. Herders do not in any case seem to regard the different human and non-human actors that populate the region as having "interests" that must be aligned or as being motivated by some variant upon economic gain, but are instead concerned with the maintenance of balanced and satisfactory relations between a variety of entities. This last point leads me to wonder whether, perhaps, decolonisation is a notion that may have relevance beyond the politics of the former Spanish dominions in the Americas. This study of animals and science in the context of the Andes might point the way to a "decolonisation" necessary in science studies that is brought to our attention by the example of the herders. Andean llama herders, in a similar manner to actor-network theorists, endow a range of non-humans with agency–the high mountains, the fertile earth, llamas, lakes and stones–and they emphasise membership of collectives above individuals, yet do not seem to have the sorts of "interests" that actor-network theorists assume. More critical scrutiny of the notions of economic rationalism, stemming from Enlightenment philosophy, that are still inherent in science studies might enable this useful perspective to become more widespread in anthropology and to move beyond the bounds of the study of science in Western societies.

Acknowledgements

The period of fieldwork upon which this paper is based was funded by the ESRC. I would like to thank the participants at "Animals and Science: Anthropological Approaches" at Manchester 2005, Brian Johnson, the organiser of a panel on medical anthropology at LASA 2006, San Juan, Puerto Rico, and the staff and students at the departmental seminar in anthropology at the University of Aberdeen for their comments and helpful suggestions when previous drafts of this paper were presented. I would also like to thank Kate Degnen, Stephanie Bunn and Julie Cruikshank for their careful readings of the paper and their thoughtful comments.

Notes

[1] I place the term vaccine in inverted commas as the Spanish term *vacuna* was used by herders. However, strictly speaking, the chemicals involved were not vaccines but broad-spectrum anti-parasitic medicines.

[2] "Racially and culturally mixed" is a common translation of the term *mestizo*. However, this gloss belies the complex genealogy and conceptual hybridity of the

term, which combines nineteenth-century ideas of race and culture with earlier notions of ancestral faith and transgression. For a thorough treatment, see de la Cadena (2005).

[3] A military metaphor employed by several actor-network theorists (see Latour (1988), Callon (1986).

[4] See for example Zulawski (2000).

[5] Most recent discussions on education in Bolivia have promoted not *interculturalidad*, but *intraculturalidad*, a model of education seeking to reaffirm indigenous identities and benefit indigenous people on their own terms (Howard 2009).

[6] For example, when a Constituent Assembly was convened to rewrite the state constitution in 2006, many hoped that state democracy could be strengthened by the incorporation into the assembly and the resulting constitution of indigenous ways of reaching consensus. However, attempts to do so were frustrated by a combination of government anxiety and right-wing opposition (Hylton and Thomson 2007:138-142).

[7] For example, Hobart (1993), Escobar (1995).

[8] For example, Apfel-Marglin *et al.* (eds.) (1998).

[9] See Bolton (2001).

[10] See Bolton (2006).

[11] Here I have italicised Spanish terms.

[12] The term *originario* was used in colonial times to designate people native to a region for purposes of taxation. Present day use of the term for people conflates this fiscal category with indigenous ideas about origin and attachment to one's place of birth (see Bolton 2007).

[13] *Samiri* – from *sama* (Aymara, "breath") + *-iri* (Aymara, "one who [gives]") (Abercrombie 1998: 357).

[14] One informant suggested in jest that they would have spoken Quechua, since stupidity is both a characteristic that Bolivians in general attribute to donkeys and that urban *mestizos* attribute to Quechua-speakers.

[15] The motivation for this may have been expediency, but most probably was insult, since the *alcalde* (mayor) of Rosario, who was responsible for the meal, had political differences with the union leadership.

[16] Although there was also a limited market in cities for dried llama meat: as Cassandra Torrico (personal communication) reminds me, urban food includes dishes such as *ch'arki kan* – dried llama meat served with maize, and sometimes potato and boiled egg.

[17] *El Potosí*, 20/11/2003: "La carne de llama era despreciada".

[18] Although Douglas (1966) and Latour write of "purity" or "purification" there is a subtle difference between their formulations. For Douglas, "pure" categories become contaminated by things that do not fit them, or matter out of place. Hence, purity precedes mixture. In Latour's formulation, the Modern Constitution involves two processes which he calls purification, or the separation of nature from culture, and translation, the proliferation of hybrids between the two. Here, purity is *simultaneous* with hybridity and hybridity enables purification. As de la Cadena (2005: 268) puts it, "there is nothing to separate if things are not mixed".

[19] Personal communication from ethnomusicologist, Henry Stobart.
[20] http://www.slowfood.com/
[21] Organization International des Épizooties, The World Organisation for Animal Health.
[22] The type of certification sought in Bolivia to date for its two FMD-free zones has been for "disease-free status where vaccination is practiced".
[23] Furthermore, Law (2004: 227) argues that the control achieved is precarious and liable to periodic and catastrophic breakdown.
[24] A similar sort of action in the case of mange was noted by sixteenth-century chronicler José de Acosta (Acosta 2002: 246).
[25] Vicuñas are wild camelids, related to llamas and alpacas, that became endangered from unregulated hunting motivated by the high monetary value of their fleece.
[26] "co-godparent"
[27] See for example http://www.cababstractsplus.org/abstracts/Abstract.aspx?AcNo =19970801007 for evidence from the Indian subcontinent. Zoonotic diseases are those that can cross the "species barrier" between animals and humans.
[28] The medication that has been used most widely used in Lípez is *Ivomec*™ a pharmaceutical developed by Merial Ltd., primarily for cattle and sheep in the US and Britain.
[29] There may be another entanglement with conservation measures here, since environmentalists have voiced concern about dwindling quantities of *yareta* in the Western highlands, blaming its decline on its use as fuel by indigenous people.
[30] My compadre, described the vaccinations as *vacunas...contra la flacura*, "Vaccinations against thinness" and as *sueros, así inyectables, eso es para la flacura*, "Serums, injectables. They're for thinness".
[31] Still-births and birth deformities are problems that scientifically trained NGO personnel attribute to in-breeding, for which they criticise the practices of indigenous herders (Bolton 2006). However, there have been reports of teratagenicity associated with the use of ivermectin in animals, and human patients are advised to avoid taking the drug while pregnant or breastfeeding. From my reading of related papers and websites, it appears that definite knowledge about the effect of the drug is uncertain and is certainly beyond the scope of this paper. See for example, http://www.drugs.com/pregnancy/ivermectin.html.
[32] Elsewhere in the Andes they are known as *pishtacos, kharisiris or ñakaqs* (Gose 1986; Rivière 1991; Salazár 1991; Canessa 2000; Weismantel 2001).

Bibliography

Abercrombie, T. A. 1998. *Pathways of Memory and Power: Ethnography and History among an Andean People*. Madison: University of Wisconsin Press.

Acosta, J. de 2002 *Natural and Moral History of the Indies.* Edited by J. Mangan. Durham, Duke University Press.

Apfel-Marglin, F. with PRATEC (eds.) 1998 *The Spirit of Regeneration: Andean Culture Confronting Western Notions of Development.* London: Zed Books.

Arnold, D. Y. and J. de D. Yapita. 2000 *El rincón de las cabezas. Luchas textuales, educación y tierras en los Andes:* La Paz, UMSA/ILCA.

Bastien, J. W. 2002 "Sucking Blood or Snatching Fat: Chagas' Disease in Bolivia", in J.Koss Chioino *et al.* (eds.), *Medical Pluralism in the Andes*. London. Routledge: 166-188.

—. 1992 *Drum and Stethoscope: Integrating Ethnomedicine and Biomedicine in Bolivia,*. Salt Lake City: University of Utah Press.

Bolton, M. 2007 "Counting Llamas and Accounting for People: Livestock, Land and Citizens in Southern Bolivia". *Sociological Review.* 55(1): 5-21.

—. 2006 "Genetic Defects or Generative Prototypes: Competing Models for Livestock Improvement in Southern Bolivia". *Journal of the Royal Anthropological Institute*. 12 (3): 531-549.

—. 2001 *Between the Ayllu and the Nation-State: Intertextuality and Ambiguities of Identity in San Pablo de Lípez.* Unpublished PhD thesis: University of St. Andrews.

Callon, M. 1986. "Some Elements of a Sociology of Translation: Domestication of the Scallops and the Fishermen of St Brieuc Bay". In J. Law (ed.), *Power, Action and Belief: a New Sociology of Knowledge?* London, Routledge: 196-223.

Canessa, A. 2000 "Fear and Loathing on the *Kharisiri* Trail: Alterity and Identity in the Andes", in *Journal of the Royal Anthropological Institute* 6(4): 705-719.

de la Cadena, M.. 2005 "Are *Mestizos* Hybrids: The Conceptual Politics of Andean Identity", Journal of Latin American Studies 37: 259-284.

—. 2000 *Indigenous Mestizos: The Politics of Race and Culture in Cuzco, Peru 1919-1991*, Durham, Duke University Press.

Dransart, P. 2002 *Earth, Water, Fleece and Fabric: An Ethnography and Archaeology of Andean Camelid Herding*, London, Routledge.

Douglas, M. 1966 *Purity and Danger*, London, Routledge and Kegan Paul.

Escobar, A. 1995 *Encountering Development: The Making and Unmaking of the Third World*, Princeton, Princeton University Press.

Fowler, M. 1999 *Medicine and Surgery of South American Camelids*, Hoboken NJ: Wiley-Blackwell.

Gose, P. 1986 "Sacrifice and Commodity Form in the Andes". *Man* (N.S.). 21: 296-310.

Harris, O. n.d. "The History of Ayllus in the Present Conjuncture: Source if Insight or Waste of Time". Discussion paper prepared for *Bolivia Forum*, Newcastle University, February 2009.

—. 1980 "The Power of Signs: Gender, Culture and the Wild in the Bolivian Andes". In M. Strathern and C. MacCormack (eds.) *Nature, Culture and Gender*. Cambridge: Cambridge University Press: 70-94.

Healy, K. 2004 "Towards an Andean Rural Development Paradigm?". *NACLA Report on the Americas* 38(3): 28-34.

—. 2001 *Llamas, Weavings and Organic Chocolates: Multicultural Grassroots Development in the Andes and Amazon of Bolivia*. Notre Dame IN: University of Notre Dame Press.

Hobart, M. 1993 *An Anthropological Critique of Development: The Growth of Ignorance*. London: Routledge.

Howard, R. 2009 "Education Reform, Indigenous Politics and Decolonization in the Bolivia of Evo Morales". *International Journal of Educational Development* 29(6): 583-593

Hornberger, N. 2000 "Bilingual Education Policy and Practice in the Andes: Ideological Paradox and Intercultural Possibility". *Anthropology and Education Quarterly* 31(2): 173-201.

Hylton, F. and S. Thomson. 2007 *Revolutionary Horizons: Past and Present in Bolivian Politics*. London: Verso.

Latour, B. 1993 *We Have Never Been Modern*. London: Harvester

—. 1988. *The Pasteurization of France*. Cambridge MA: Harvard University Press.

Law, J. 2004 *Disaster in Agriculture: or Foot and Mouth Mobilities*. Centre for Science Studies, Lancaster University: http://www.comp.lancs.ac.uk/sociology/papers/law-disaster-mobilities-foot-and-mouth.pdf.

—. 1986 "On Power and its Tactics: A View from the Sociology of Science". *The Sociological Review,* 34: 1-38.

Leach, J. 2007 "Creativity, Subjectivity and the Dynamic of Possessive Individualism". In E. Hallam and T. Ingold (eds.), *Creativity and Cultural Improvisation*. Oxford, Berg: 99-116.

PIEB n.d. *La interculturalidad: Aproximaciones, alcances y rupturas*. http://www.pieb.org/campesinosenlauniversidad/contents/EA%20intercult.pdf. Accessed 17.03.2009.

Rivière, G. 1991 "*Lik'ichiri* y *kharisiri*… A propósito de las representaciones del 'otro' en la sociedad aymara". *Bulletin Institut Français des Études Andines*, 20 (1): 23-40.

Salazár, C. 1991 "El *pishtaku* entre campesinos y mineros de Huancavelica". *Bulletin Institut Français des Études Andines*, 20(1): 7-22.

Weismantel, M. 2001 *Cholas and Pishtacos: Stories of Race and Sex in the Andes*. Chicago: University of Chicago Press.

Zulawski, A. 2000 "Hygiene and the Indian Problem: Ethnicity and Medicine in Bolivia 1910-1920". *Latin American Research Review*. 25(2): 107-129.

CHAPTER EIGHT

HOW BIOTECHNOLOGY MAKES HUMAN KINSHIP WITH WILDLIFE VISIBLE

BEN CAMPBELL

Relationships with animals and interactions of humans with the non-human world are frequently integral to ethnographic studies. Through close attention to livestock herding, to hunter or horticultural species interactions, and to symbolic affinities of social groups with floral and faunal familiars, anthropologists have found illustrative and even systematic points of explanatory context and comparison for talking about human sociality (Willis 1994; Rival 1998; Knight 2004.). Kinship in particular is hard to discuss among pastoralist peoples without due consideration to hoof and horn. A Nuer idiom for one of their nodal lineage segments is not a discretely human referent but 'the smouldering dung heap' (Evans-Pritchard 1950: 362). When we talk of kinship in contemporary European society, it is most unlikely that animals will figure prominently. Place will perhaps provide some non-human context, and in making places into 'home' domestic animals might feature, but on the whole discussion of kinship will be about circuits of human relatedness and the making of proximity and difference in the instrumental and affective bordering of private and public. The presence of the non-human can be either effectively ignored, or consigned to some contrastive realm, devoid of intention and relational concern, as a resource or a backdrop to distinctly human concerns. We do not expect to find kinship 'out there' between human and non-human realms.

Such is the cosmology of modern naturalism according to Descola (2005). He systematically analyses how the notion of an objective material world of the non-human, singularly knowable through science, belongs in a structure that simultaneously creates a domain of the subjectively organised individual human actor, and multiple worlds of cultural meaning. For Descola, this places our dominant orientation to the non-human at odds with other societies' animistic, totemic or analogical cosmologies. For us, nature can be known through empirical science, and the life-worlds of beasts and botanical communities obey nothing but mute laws of food resource availability and competitive advantage. By contrast, the notion of human exceptionalism presents people's behaviour as uniquely configured through cultural meanings, and the attribution of conscious intention and deliberative agency is restricted to humans alone. Building on Haraway (1997), Franklin (2001), and Latour (2004), my interest here in genetically modified (GM) crops is in how they confound our ideas of what constitutes an object, and in how parts of previous wholes are distributed in recombinant associations of stuff and purpose, for which the idiom of kinship can provide a useful framework for tracing connections. My strategy in this chapter is to suggest that this neat modernist division has been destabilised by biotechnology, and that the new frontiers of knowledge and practice have called up the ghosts of relationships between humans and the non-human, camouflaged by naturalist cosmology, and which in the scheme of comparative anthropology could be discussed as a language of kinship.

Scientific knowledge of human intervention in other species' habitats, and their survival prospects, provides material for thinking anthropologically about the compartmentalisation of scientific knowledge as objective fact, and its epistemic interface with relational forms of knowledge and value.[1] Biotechnology extends the frontiers of possibility for human agency to redesign interactions with life processes, and the terms of advantage in pursuit of goals such as efficiency in food production. The question I ask is whether agricultural biotechnology's intensified application of rational calculation over the non-human world has pushed us up against a significant cosmological threshold. What is at issue is the post-enlightenment division of the human and the non-human, that reads modern social relations as biologically restricted to inter-human affairs. With the biological discreteness of human and other organisms being rendered capable of re-engineering, has human exceptionalism had its day in constituting the terms of the social?

Congruent with Schneider's analysis that Euro-American kinship ideas are ruled by the principle of biogenetic connection, kinship with non-humans has come to be known through degrees of shared evolutionary DNA (Jones 2000). Though our proportional genomic relatedness with chimpanzees has emerged as much closer than many had imagined, the separation of humans from non-humans is seen as a biological given. Kinship with primates can only be spoken of in terms actual affinities broken off some 6 million or so years ago. So runs the 'naturalist' world view of modern times in which materiality follows scientifically knowable universal laws, and human affairs of a distinctly non-material kind constitute separate domains of cultural relativity (Descola 2005). Indigenous people may be accredited with ideas of a 'kinship with the natural world', for example where a forest may be seen as parent to its human population (Bird-David 1993), or varieties of rice are asserted to be offspring of a metamorphosed human ancestor (Novellino 2007), but from a modernist point of view this is incontestably a metaphorical projection of human concerns onto the natural.

The subject-object separateness of the human from the non-human came to be, as Descola puts it, a cosmological transformation in which humanity is exceptionally endowed as a cultural animal with unique intentional agency towards the world out there. Inter-subjective communication through cultural tools contrasts to a domain of brute material non-humanity. In this framing there can be no kinship across, except as a conceit. End of story.

The argument proposed here is that the framing described above would restrict the possibilities for looking at the creative instability of 'naturalist' dynamics in associations of the human with the non-human. If kinship for comparative purposes is rather understood not through shared biogenetic substance, but as a technology for the ordering of solidarity, desire and exchange, at a minimal and necessary level, the frame for analysing human interaction with the non-human is no longer ethnocentrically constrained. By looking at events and instances where modernity's great divide becomes existentially problematic, and its epistemology no longer compels but disturbs, the ground may be on the move. Descola is persuasive in tying up relations with the non-human in terms of extractive resource exploitation, on the one hand, and the ethic of biodiversity conservation on the other. They are variants of the same mythical template; a singular cosmology of division between culturally relative subjectivities and objective knowledge of the material world. There are indeed individuals who happily shift between transnational corporations that 'manage' both (from oil companies to WWF); poachers become gamekeepers without turning the world upside down. However, on a number of levels within the debate over genetically modified organisms in the UK there was cause for questioning Descola's argument that the naturalist cosmology has not encountered an alternative.

In this paper I focus on media reporting of a series of public debates on the introduction of genetically modified organisms (GMOs) into the foodchain in Britain, organised by the UK government and held in 2003, and reporting of the results of farm-scale evaluations of GM crops published in the same year. Before leading into the discussions and reportage of GM-related controversy by the media, let me highlight that the issues relating to the agricultural biotechnology debate in the UK of most concern in this chapter are:

- The way power was vested in scientific scrutiny to speak about agricultural technology as human intervention in the non-human world;
- The translation from fact to meaning in predictive measurement of human intervention on non-human populations, that provided the basis on which deliberation between desired options were to be decided in public arenas;
- The reception of the scientific data in a moment when the stakes were perceived as, for want of an alternative term, 'sacrificial' at a species-level in the implications for the creatures of the countryside;
- And lastly, as an instance where in quasi-totemic, or analogical, fashion the non-human world came to stand as a domain carrying the burden of inter-human struggles.

What this points to is a threshold in the extent to which the non-human could be acceptably treated as a resource submitted to human will and design. Science was to be assigned a key role for the public debate, and this was contentious from the outset as the commercial and geo-political context of GMOs in the hands of biotechnology corporations was more likely to be the elephant standing in the room, rather than the quality of the food on the table. Official, government-sponsored scientific survey reporting was expected by many observers to be partial to biotechnology, and the industry followed the public-as-knowledge-deficient narrative, offering a new dawn of production efficiency through the wonders of gene manipulation as an improvement over the natural properties of conventional crops. Government included a spectrum of opinion, and positioned itself in a mediating position between commercial interests and consumer activists. However, science was rhetorically deployed by both sides of the pro- and anti-GM divide, and a context for public consultation and ethical deliberation was provided through regional meetings, to air components complementary to the science, to allow for validation and embedding of scientific data into publicly accountable decision making processes. If things were as hard and fast as Descola claims, that the social and the voicing of relational community belongs to an internal human domain, the images and phrasings that were notable in newspaper treatments as I discuss below could only be seen as an exotic aberration. As it is, the degree of reflexive anticipation about GM technology's capacity to silence the sounds of the countryside, that would make it actually 'mute', in the sense of inaudible, rather than just communicatively inconsequential, suggests a decisive moment was reached in the pursuit of human mastery over the non-human.

The Structure of Debate and Scientific Input

Responding to widespread alarm in a number of European countries in 1996 over the import of GM soya from the USA in shipments that did not segregate GM from non-GM produce, and the arrival of foods (e.g. tomato paste) with GM content on supermarket shelves (Bauer *et al.* 2002), a number of reports ('Crops on Trial' by the AEBC, 'The politics of GM' (ESRC 1999)) had proposed a public debate over the introduction of GMOs into the food chain. The Agriculture and Environment Biotechnology Commission (AEBC) was given the task of co-ordinating public debate. In a letter to Secretary of State for the Environment, Food and Rural Affairs, Margaret Beckett, on 26th April 2002, the commission stated:

> Our aim is to promote a debate that would be an important example of public participation in discussion of scientific issues. The programme we are recommending… has also benefited from the growing understanding of the experience of other countries such as Denmark and the Netherlands in developing processes for stimulating constructive public discussion of the social and ethical implications of new developments in science and technology. We would hope that lessons could be learned from implementation of the programme we are recommending that would be of wider applicability to other areas of science-based policy-making. (Point 9).[2]

In the designedly neutral functionality of this dialogue, the explicit government position was that it should be at arm's length from the debate. The debate may well not have occurred had it not been for the presence of certain sceptics among the occupants of ministerial positions of the time.[3] Against the view that GM represented a new way forward for farming efficiency, and innovation, and promised profits in the otherwise bleak food sector there were many reasons people gave for opposing the introduction of GMOs. These ranged from 'tampering with nature', and lack of information for consumers about the traceability of products, to concern for multinational control of the world food supply.

In terms of WTO trading agreements, the only legal basis for a 'derogation' not to import or grow GMOs was evidence of harm to human health or the environment. The British government's scientific advice suggested little evidence to question health implications, and as a complement to the parallel public opinion consultation exercise held in 2003, a decision was made to vest scientific examination of the crops in the area of environmental impact. 'Farm Scale Evaluations' (FSE) were conducted on over two hundred sites across the country to look at the effects on the plants, insects and birds found in field locations, of growing crops modified for herbicide resistance. In the initial formulation of the proposed public debate, the scientific evidence was designed to be available prior to the regional public meetings and the survey of the general public's opinions on GM. To the dismay of many on the steering committee of the AEBC, the Farm Scale Evaluations were eventually scheduled for release only after the public debate. The results were published on 16th October 2003 and were widely reported and discussed by the press on front pages and in editorial comment.

Was the media reporting congruent with the naturalist cosmology of a non-human environment of 'mute' entities, whose life processes can be revealed by science? On one hand a clear reverence for the revealed 'objective' truths of the findings was apparent, and the extent and systematicity of the ecological survey impressed the journalists. But I argue that latent social readings were also detectable. These invoked the missing relationship of people (beyond farmers) to the agricultural environment in the framing of science. For example, the newspapers gave an iconic status to birds, referred to by their evocative 'common' names, picking up on the use of birds as indicator species for the effects of changes in crop management. Furthermore the interdependence of life forms in field ecologies, from springtails to skylarks, under threat from dosings of weedkiller to which the GM crops were resistant, by implication included humans. In the context of a managed public debate in which social, political and cultural arguments over GM technology were sidestepped by the weight of scientific authority, the birds, weeds and mini-beasts *stood for human presence in a quasi-analogical way*. All the weight of social concerns about GMOs was now hanging in the environmental basket.

To substantiate my claims, I will concentrate on two themes of the media reporting of the results: the picture of biodiversity in field ecology that emerged from the trials, and the consideration given to the desirability to people and society of further intensification of farming.[4].

Many newspapers presented selected factual findings adjacent to commentary.[5] *The Daily Telegraph* noted that

> GM oilseed rape and GM beet damaged the environment. There were fewer butterflies, bees and invertebrates because there were 80% fewer weeds and seeds for them to eat.[6]

The Guardian headlined a section "Birds and bees: how wildlife suffered", itemising the different outcomes for a range of species under GM and conventional farming.[7] The springtail, a small wingless arthropod, was one of the few species that seemed to benefit under GM oilseed rape. This was because the timing of weedkiller application increased the amount of decaying plant matter available to it. (GM scientific evaluation p.33)

The *Independent* front page declared that the results of the Farm Scale Evaluations confirmed:

> …conservationists' concerns that the GM crops scheduled for growth in Britain would mean yet another blow for the insects, flowers and birds that have been decimated by more than 30 years of intensive farming[8]

For *The Times* countryside editor the results painted "a grim picture of a landscape denuded of many farmland birds, butterflies, insects, and common field plants." A quote was included from Mark Avery representing the Royal Society for the Protection of Birds, that GM crops could mean "the final nail in the coffin for some species".[9]

The tabloid *Daily Mail* devoted a two-page spread to the results, and declared "farming the so-called Frankenstein crops risks creating a biological desert by wiping out wild plants, butterflies, bees and birds."[10] *The Guardian's* leader pronounced that the further deprivation of habitat and food from birds and animals would be "an irreplaceable loss to the countryside which once teemed with the sights and sounds of creatures on and above the ground". The editorial in *The Independent* noted the finding that "Fewer weeds means fewer – and fewer varieties of – insects, and that in turn means fewer corn buntings, skylarks and yellowhammers".[11]

The Guardian reported that governmental responses "were cautious", but the Environment Minister Elliott Morley was quoted: "GM crops had severe implications for birds". Paul Rylott from the industry body promoting GM argued against the chorus of doom that "scaremongering is not supported by the facts" and claims that the crops would "wipe out wildlife" were unfounded.[12]

What, it might be asked at this stage, has all this to do with kinship? Consider the phrase "severe implications for birds". In the modern naturalist cosmology, the creatures themselves will not deliberate the consequences of GM oilseed rape, beet and maize in an assembly of feathered representatives (the fact that images in the likeness of Beatrice Potter may come to mind merely emphasises the fanciful quality of such a proposition). It is humans who are implicated and signified in the knowledge about likely outcomes for avian life of intensified applications of weedkiller in the fields. It is the human senses that register 'irreplaceable loss', 'grim picture', and the successional deprivations along the food chain from weeds through insects to birds. It is to human ears and eyes that disappearing sights and sounds are expected to matter, starving chicks with gaping mouths notwithstanding. Radcliffe-Brown's totemic question "Why all these birds?" (1951: 17) deserves to be legitimately repeated here, and as Lévi-Strauss put it "The connexion is not arbitrary, nor is it a relation of contiguity" (1962: 147). What kinds of cross-species solidarities, desires and exchanges are raised up in the brute demonstration of impending biodiversity decline will be considered in due course.

The desirability for British field ecology to be dedicated to maximised efficiency in farming practices also came into question. The Mirror's dramatic headlines emphasised the choice on offer, borrowing Rachel Carson's book title "Silent Spring", and a sub-heading pitched the contesting options for the use of the countryside "Birds and Bees: technology v wildlife". The Independent's editorial writer presented a scenario of clear alternatives:

> …the choice comes down to enhanced yields versus reduced biodiversity. … there has been a growing recognition in recent decades that farming policy should be tending in the opposite direction to that offered by today's genetically modified crops. The movement should be away from intensive farming and towards the preservation of biodiversity. … the environmental costs of modern farming methods are too high.[13]

'The preservation of biodiversity' is an appeal familiar within the naturalist cosmology of modernity, that resonates in nineteenth century projects of safeguarding areas of 'wilderness' (Cronon 1996; Olwig 1993). But the GM debate revealed a distinct contrast in how the non-human is accommodated on different sides of the Atlantic. In the US, setting aside large areas for wilderness enables other areas to be hyper-intensively cultivated. In the UK's more space-restricted, 'small and windy island' as the Public Debate report put it, the accommodation of wildlife and farming is as it were more 'face-to-face', more relational.

The Guardian's account of the trial results set the government's policies for promoting biotechnology against other pledges it had made concerning protection of the environment. Chief among these was the aim of reversing the loss of birdlife in the countryside, which the government had identified as a 'quality of life' indicator (presumably intending *the quality of human life*). Here the classic-modern productionist logic of agricultural efficiency met a late-modern post-productionist counter-valuation of the countryside as a place of locality, lifestyle and leisure consumption.

The *New Scientist* journalist Andy Coghlan observed that farmers in the US and Australia use broad-spectrum weedkillers resulting in:

> …fields sterilised of everything except the crop. Farmland there is purely for business, and if people want to see wildlife, they can visit national parks. But in Europe, farmland is used for leisure as well as producing food, and conservationists want farmers to be kinder to what wildlife remains.[14]

Coghlan added the UK has witnessed a "catastrophic decline" in a number of bird species in the last fifty years with modern agriculture's expansion of field sizes at the expense of hedgerows, made possible by the use of agro-chemicals, and the increase in winter growing patterns, that remove fallow sequences.

Counter arguments in support of GM as an efficient direction for British farming were presented in *The Daily Telegraph*. Its editorial argued that anti-GM 'feeling' is simply motivated by "fear of the new".[15] In its coverage of the pro-GM side, *The Independent* included Paul Ryllott of the Agriculture Biotechnology Council claiming, against the tide, that "GM crops are more flexible and can enhance biodiversity". Key to this argument was the finding that of the crops being trialled, GM maize was less damaging to wildlife than cultivation of 'conventional' maize, for which the especially toxic weedkiller atrazine was used.

The Daily Mail also quoted environment minister Elliot Morley restating the legally determined position regarding derogation for the new crops: "GM crops can only be grown if they get consent. Whether they get consent depends on whether there are environmental impacts." 'Consent' is not here a matter of public approval, but a scientific underwriting to meet legal formalities. What is of particular interest in this respect is how the social dimension of technology assessment (made up of values, beliefs, interest groups etc.) becomes compartmentalised as separate from 'environmental' factors (demonstrable through scientific fact). The effects of this boundary-making are powerful in giving priority to scientific realism, establishing a foundational domain of nature and biology as a discrete and more dependable kind of objective evidence compared to subjective public opinion.

Fields of Deliberation

The reports on GM impact released by The Department of the Environment, Food and Rural Affairs (DEFRA) presented divergent scenarios for different bird species in the scientific modelling of different management options for GM crops. Bill Sutherland's summary points to variables of rotational cultivation, options of leaving fields fallow over winter, and the percentage provision of uncultivated field boundaries, as all contributing to very different kinds of outcome for different species. Even the dietary significance to birds of various invertebrates is pointed up as an area of little knowledge, which could have a bearing on herbicide-tolerant (GMHT) crop introduction.

Modelling of the FSE results delivered scenarios for assessing combinations of land use strategies that make GM cultivation less harmful to certain birds. Thus Sutherland considers methods for the 'amelioration of GMHT crop use':

> It has been suggested that such is the control of weeds in GMHT crops, that there is less need for weed control when the crop is not in the ground. It is possible, therefore, that the effects of GMHT crop use could be ameliorated by the increased prevalence of land left uncultivated over winter. The effects of the simultaneous switch from a 5% to a 10% stubble regime and the introduction of GMHT crops on the population densities of farmland birds predicted benefits of the increased prevalence of uncultivated land to curl buntings which far outweigh the damage done by the introduction of GMHT crops. However, the same is not true for the skylarks, which decline as precipitously even when stubble frequency increases. (DEFRA 2003: 20, point 12.3.4)

These scenarios present a knowable non-human world 'out there' that will respond to new farming technologies in ways that can be predicted. Is this a case of an incremental alienation from a 'nature' no longer autonomous and treated as a resource made ever-more amenable for human purpose as Descola would have it? The various scenarios of science come across as value-free data enabling deliberation and contention between competing sets of values, in a well-informed process of taking variables into account. The question arises of how any extra-human relatedness could be made visible when the vegetal processes of the soil and the creatures living from them are made negotiable and become a matter of choice? However, contrary to the representation above of science's relation to society, an alternative ecology of knowledge informed by comparative anthropology is possible in order to appreciate the mythical qualities of the science/politics divide, and its epistemic boundary-making.

Nowotny *et al.* (2001) provide a cogent analysis of the conditions in which science's socio-political life has transformed in the last decades. With the lines between science, politics and commerce increasingly blurred, and the old white-coated authority of the laboratory contending with the claims of 'non-scientists' to kinds of expertise that cannot be straightforwardly over-ruled, arenas for democratic scientific practice have come into being. The UK's 'GM Nation?' debate belongs firmly in this process, but with its separation of environmental science from public dialogue, the possibilities for engaging in debate over the science were strongly stage-managed. Indeed, substantive public engagement with the science was precluded by the scheduling of the public debate prior to the FSE trials being made public.

There have been many other examples of fora or 'agora' for such discussion, and it is pertinent that one of the very first was a 'consensus conference' held in the Netherlands on the topic of genetically modified animals (Mohr 2002: 147; Klüver 1995; Mayer 1997). (The outcome was in fact not consensus, but a greater division of opinion among participants after the process, than had existed before). There lies a strongly rationalist premise behind these attempts at dialogue over scientific ethics and politics, that aims to reduce the information deficit among participants, whether of a scientific kind, or to enable different protagonists to hear 'where people are coming from'. Many reports on the outcomes of these deliberative events are not optimistic for the triumph of reason. I mention attending one of the afternoon-length UK fora for public consultation below. Alison Mohr's account of Australia's public deliberation on GMOs over three days in 1999 highlights the difficulties in achieving any genuine exchange in such events. She catalogues the unresponsiveness of spokespersons from Monsanto to questions from panellists, and the selection of certain kinds of participant:

> A lay panellist was extremely disappointed by the identification early in the process, by the facilitator and publicist, of the 'housewife', 'yuppie' and 'pretty young thing' to act as the public face of the lay panel. As a consequence, she felt that "having the media there reinforced that some people's views are more important and more valid than others" (Mohr 2002: 162-3).

When a panellist asked the Monsanto representative if its products were safe to humans, the representative admitted to not having the information. The panellist later demanded an answer from the representative outside the chamber. The next day the answer was given in informal exchange, but the panellist insisted the representative told the whole gathering, saying "I know the answer but the people in there don't know the answer".

> Another lay panel member agreed with the perception that [the Monsanto representative] "failed to address anything that we asked him. He had this half-hour window of opportunity to set the record straight on Monsanto and at least present their case a little more favourably and I felt he failed to do that". The panellist was particularly disappointed as he "wasn't anti-Monsanto before [he] attended the conference but during the course of the conference [he] started to have doubts about the validity of [the Monsanto representative's] position and statements because he just seemed to be practising avoidance strategies". An expert speaker concurred, labelling [the Monsanto representative's] presentation "disappointing, there was a lack of honesty, candour, and a lack of a genuine attempt to answer the question". (Mohr 2002: 169)

One expert commented the whole exercise had been much publicised, but there was no effective means for bringing the lay panel's findings and recommendations to the attention of decision-makers and policy regulators.[16]

It might be thought surprising that Monsanto and other biotechnology advocates did not show more postmodern scientific reflexivity in attempting to make the first generation of GMOs publicly acceptable, and were so uncomfortable in having to do more than just roll out experts in economics and science. This could be confirmation of the resilience of the naturalist cosmology among the very people who are arguably dismantling the divide between humans and non-humans by DNA transfer of human intentions into modified organisms (designing crops to respond to specific environmental conditions for purposes of productive benefit to the farmer). By pushing beyond the natural limits of crop varieties for new possibilities of profit from agriculture, the productionist rationale for innovation extends the cosmology of human exceptionalism to alter the non-human world at will.

Gottweiss analyses how the biotechnology industry has defined the risks it generates as controlled by the technologies that produced their possibility. Regulation became 'an expert enclosure' (1998: 328). A micro-politics of meaning emerged in the discursive production of genetic engineering in multiple sites, from laboratories to parliaments - creating, structuring, and prioritizing the range of possibilities available for policymaking by government agencies. "[T]his micro-politics of meaning relied on policy narratives that established relationships among economic growth, molecular biology, recombinant DNA's hazards, technologies, collective identity, and international economic competitiveness" (*Ibid*.331). As in the Australian hearing reported by Mohr, the fall-back to short circuits of supportive rationales for the economic and scientific case in favour of GMOs avoids an engagement with the open-ended directions that public discourses bring to the table in specific country contexts, with their different 'cultural' relationships to food and to the patterns of the use of science in food policy.

The power of biotechnology depends on a particular misrecognition of its own political agency in a cosmology that separates the science of nature and politics. The world of nature cannot speak for itself, while politics is only talk. For Latour these "continue to be seen as two completely unrelated sets…We are still expecting our salvation to come from a double assembly, only one of whose houses is called politics, while the other one simply and modestly declares its determination to define matters of facts" (2004: 31). What he calls 'the Myth of the Cave' distinguishes two houses "one of which chatters away in ignorance while the other has knowledge but does not speak, the two being connected by a narrow corridor through which, by a miraculous and double conversion, minds that are scientific enough to make things speak and political enough to silence humans go back and forth" (2004: 92).

The public media consumption of knowledge about threats to farmland wildlife seemed to respect the narrative power of scientific expertise conveyed by descriptions of dynamics in the natural food chain. It was engaging as a story of another world, that humans could behold and appreciate at a relation of respectful distance. If the lay-expert divide appeared to be credible in a moment of awe displayed in front of the extent and content of what the research revealed, the scientific frame did not for long hold the discursive bounds of the debate. In the *Guardian* of February 20th 2004, Clare Devereux of the Five Year Freeze coalition commented:

> "The government just does not get it. The public is way ahead in understanding that agricultural biotechnology is about a lot more than just the science. It is about livelihoods, choice, culture, the biodiversity of our landscape, the survival of small farmers – and GM crops could potentially threaten all of these"[17]

The ring-fenced modern cosmology of naturalism (Descola 2005), or as Latour (2004) puts it the 'bicameral constitution' of knowledge, compartmentalises environmental truth, setting it apart from the relationships of social and political domains. However, this was to let loose recalcitrantly symbolic inflections of the 'environmental' to imply human significance, and refracted narratives of the social and political. For public consultation to be managed in orderly fashion as an interface between various kinds of technical expertise and popular participation, science, economics and ethics had to be kept initially separate, only subsequently to find dialogue. For many who came to the regional 'GM Nation?' meeting at Harrogate I attended, this containment was almost unbearable, and only at the end of the afternoon's session did a more vernacular, less managed, set of demotic exchanges take place in the closing minutes. In the voluble interjections between enthusiastic farmers and activists of many hues, "the empire of passions" spoke to the "indisputable order of Science" (Latour 2004: 88).

Managing Orders of Truth: Kinships Old and New

The newspapers dramatised the facts and projections for the loss of mini-beasts and their predators, but did not on the whole put the relationship of field ecology and society on the table, other than as a choice about economic or biodiversity values. The science about nature was, as it were, imagined to speak for itself, or to stand as a view on what was happening 'out there', with the idea of people's relationship to the landscape contained in a pregnant silence, and interpretation of the facts was left as a secondary process of a different epistemological order. This reflects the distance of protected authenticity by which Descola (1996, 2005) has characterised Western relationships of 'naturalism' towards the environment. Science was very effectively kept apart from any visibility of emotional or other relational connections people might have with the agricultural landscape, skylarks, or corn buntings. Even mention of the loss of bird species as a negative 'quality of life' indicator sounds distant, arbitrary, and abstract – a devised correlation, devoid of substantive content that would convey the qualities of human-environmental interaction. The environment was there to be seen and taken in as a totality prior to techno-intervention, but entirely missing the historical element of human presence, that for millennia has acted on the species links in the food chain.

Given the cool, dispassionate, facticity of the media's reporting on the Farm Scale Evaluations, it came as a relief to find eventually a human dimension of interactive and personal presence relating to this threatened ecology. It came in the figure of Les Firbank, the scientific leader of the

FSEs. He was interviewed in the *New Scientist* (25.10.03 p.46) and gave a view on a human connection with the British farming landscape, more than any other report on the science had managed. Raised on a small farm in West Yorkshire, and having gained his PhD with a study of the population dynamics of the corncockle (an arable weed), he described his family's direct experience of ecological change undergoing the intensification of modern farming methods:

> "I can remember a period when I was a kid in Yorkshire when we had cowslips all over the farm. By the time I was 10 most of them had gone. We thought it was just one of those things. We had no idea it was because of the way we were managing the land. We had put too much manure on it. ***As a family we didn't want to do any harm***. That issue – how to balance farming and wildlife – has always driven my research interests." (my emphasis)

So at last a human dimension involving elements of a very conventional British 'kinship' set in natural context – a family, with a livelihood of interaction with fields, animals and plants - was revealed to motivate the chief scientist. This placed his reflexive research practice in a relational context of human conviviality with the non-human. The Scotsman newspaper echoed the possibility of a conscientious relationship to the farming landscape by using an old-fashioned term of kinship with the land – 'good husbandry'.

As anthropologists have revisited the scope for the study of kinship in post-Schneiderian terms, exploring the productive effects of the discipline having said 'no' to the old naturalism of its object (Weston 2001: 149), attention has focused on examining how "the very demarcations, enmities and solidarities that produce family ties" (Weston 2001: 153) are produced and reified at some remove but analogous to and implied in other orders of being, collectivity and belonging. Feeley-Harnik (2001) has taken this line in tracing the background of L. H. Morgan's kinship analysis that stemmed from concerns for breeding livestock in nineteenth-century USA, and the activity of home-making observed among species like the beaver. Haraway, meanwhile, finds kinship an appropriately elastic concept for teasing out the implications of innovative order and relation that transgenic organisms bring about. She remarks they "simultaneously fit into well established taxonomic and evolutionary discourses and also blast widely understood senses of natural limit." Further, she pursues the substantial power, and hybrid agency of recombinant DNA life forms in commenting that "refiguration of the kinship between different orders of life, the generative splicing of synthetic DNA and money produces promising genetic fruit" (quoted in Franklin 2001: 313-4).

Kinship is no longer contained by genealogy, or shared DNA, but moves in unpredictable pathways of fertile conjuncture, bringing familiar co-residents of home-places to contemplate unprecedented kinds of association – even with birds. Some might say this is moving the goalposts for defining what constitutes 'kinship', but if a critical ethnographic contemplation is made of how post-natural kinship in Europe can be made socially insightful, it will not be merely in the 'strategic naturalisations' (Weston 2001) of technologically facilitated human fertility, but also in the worlds of intended relational closeness, through which people live their immediately significant quotidian being. Non-human participants cannot not be excluded *a priori* from the analytical field of kinship except by uncritical acceptance of the hegemonic folk model of science, and its privileging of shared bio-genetic substance for defining degrees of relatedness. Comparative anthropology will find no new surprise here, and it was Radcliffe-Brown who pondered of indigenous discourses on human twins, "why all these birds?" In my research in the Himalayas (Campbell 2004), the models of pastoralist nurture and predatory hunting infuse how human kinship is both symbolically constituted and effectively lived in daily activity that constantly blur categorical distinctions of human and non-human.

Science, Relational Subjectivity and Non-Humans

Unlike the pre-modern analogism of correspondences (Descola 2005), modernist analogy works through the non-human as other, and the exceptional status awarded to humanity turns out to be not so much a barrier as a bridge that can be crossed back and forth. Science is even a key means for undertaking this passage. Barbara Smuts has shown this with her work on primate behaviour, and the relational attendance she received from the Tanzanian baboons who were not interested in her as a detached scientist, but as a person in their society (Smuts in Coetzee 1999). The drive for knowledge of the other is simultaneously a drawing-in, and a finding of connection and push to identify with that otherness, as Argyrou (2005) in effect argues with the same modern cognitive structure, that Descola otherwise consigns to separation. It is the forced polarity of scientific objectivity and conscious interiority that most often stands in the way of 'holistic' anthropological understanding.

"You cannot win an emotional argument with science" (*Farmer's Guardian*, 17 Oct 03 p.11 'The view from Brussels') was the stance of biotechnology defenders policing the epistemic frontier of experts confronted by ignorance. Kay Milton has effectively pushed this kind of statement off its plinth. She identifies emotion as a driving motivation for scientific enquiry, not antithetical to it. In response to Luhrmann's question of why a flourishing of nature-drenched mythology has come about in recent times, she responds "it is because we are increasingly aware of nature's responsive relatedness to our actions. Ironically, it is science which, for the most part, seeks to understand nature as an impersonal system, that has created this awareness" (Milton 2002: 51). The opposition between emotion and scientific rationality has to be seen as a myth of capitalism, that serves to silence the voices of women, and promote the treatment of people and nature as resources (Milton 2002: 136, 150).

The mutual attentiveness of people to the environment in modern times is hard to square with Descola's naturalist cosmology, of which the prime exemplars tend to be philosophers and scientists, missing entirely the habitus of ordinary folk. Feeley-Harnik (2007: 173) argues that there has long been in Britain an association in which birds have acted as 'salient markers' of transforming ecologies, evident in Darwin's curiosity for doves and finches. The crop trials were an occasion for a new application of this historical relationship of mutual attendance between people and birds, but now in terms of the language of technological choice, or as the FSE trial coordinator expressed it, the deliberative 'share' between agriculture and biodiversity. As the public debate ('GM Nation?') was purposely conducted prior to the release of the scientific report, the social, cultural, and political economic consequences of GM farming were attributed unequal value status as compared to the authority of science. My argument is that in this scenario the creatures of farmland bio-diversity therefore became invested with, or substituted for, public voices to make a sociality of life, photosynthesis, and controlled ecological competitiveness acquire a narrative of moral-political judgement about changing times. The analogical mode of relation can survive in the pointed refusal of the configuration of public-science dialogue to indulge sentient empathy, or accord rights to other species for co-existence.

My argument is that despite Descola's naturalist cosmology, modern people have (1) practices of family and relational identity that confound the brute segregation of human and non-human, and (2) the prospect of fewer birds showed a projected co-citizenship (contra Descola 2005: 542) with non-human others in a post-exceptional (Haraway 2008) image. This is not so much the 'lost innocence' that he speaks of, but a new visualisation of relationships with entities that drags modernism's mute nature from its condition of silence by an all-too-human intervention, through which contracts with non-humans almost have to become envisageable. Descola is, on the other hand, wholly accurate in respect to the *formal* naturalist cosmology of separating knowledge of physical unity of the world from unique interiorities of human diversity. Indeed, Firbank deflected from being drawn into interpreting the FSE results and left the handling of information about GM impact on wildlife to a thoroughly human, political 'judgement call' (*New Scientist* 25.10.03 p.46).

Latour, also unconvinced by the opposition of science and irrationality, argues:

> "Whose fault is it if Science is threatened by the rise of the irrational? It is the fault of those who have invented this implausible Constitution that makes the system so fragile that a grain of sand would block it; it is not the fault of the era, which is spilling out of this ill-conceived system on all sides" (2004: 92)

Yet Latour's new constitution carries the template of public politics for new associations with non-humans rather than the compelling intimacies of 'kinship' of people's vernacular practice. This is why I have framed the chapter through the term kinship, and intend a reappraisal of late modern developments that enfold biotechnology's relation to the human and non-human in an account of social change.

Gottweis' account of biotechnology's genesis and discursive conditions of counter response is relevant here: "The fields of conflicts broadened from political struggles in the strict sense...to struggles centred on the transformation of personal and collective life, which now were, to a considerable degree, subject to the intervention of new technologies....By radically questioning the prevailing codes and descriptions of reality, social movements challenged the dominant logic on symbolic grounds" (1998: 239). Against this background "[t]he critique of genetic engineering must be seen as a part of this cultural struggle over the power to name and thus shape the world" (*Ibid*: 239).

GMOs can be seen as Escobar's (1999) techno-nature, but trapped in a productionist logic of nature as mute resource, against which non-productionist valuations of the countryside and relations with the non-human contended. Alternative techno-natures include projects for 'wild-ing', in which humanly desired landscapes of combined leisure space, biodiversity interest, and watershed functionality are held together by remotely sensed monitoring of soil conditions, water quality and species composition. The prolonged warfare in the countryside that intensive farming involved was a complicity of class dynamics particular to the welfare state, based on a productionist model of the non-human world as a resource. Post-industrial, rurally relocated urbanites with leisure and spreading motorway systems in the UK were disturbed by the severity of profit-oriented commoditisation in the countryside. In contrast to post-war industrial farming oriented to maximised yield, organic approaches to food cultivation, niche markets of artisanal food products, and the appreciation of the countryside as an environment for consumption rather than productive orientations (now served more cheaply by global food imports) changed how and what the public expected from the landscape. Rather than object-utility, qualities of subjective relation were discovered,[18] and valuation of birdlife, animal welfare, and habitat conservation displaced the hegemony of industrial farming that experienced its series of crises in Common Agricultural Policy subsidy reduction, mad cow, and foot and mouth. Peering out from behind the mask of modern productionist relationships to the British countryside, there emerged a set of commitments and common neighbourly interests that drew on old practices of identification, and possibly augmented by Disney-esque sentiment, to claim a kinship with non human survivors of the industrial age in rural Britain.

Conclusion

The first generation of agricultural GMOs were the issue of a classic productionist logic of efficiency through technology, calculable cost and benefit and containable risk. The countryside was configured as a resource for human purpose. By the mid-1990s a very different set of parameters were socially ascendant in the post-productionist logics of biodiversity, niche farming for quality rather than quantity, and cultural landscapes managed for tourism, with their quite different connections of subjectivity and the non-human. By contrast, GMOs arrive containing a programmed genealogical kinship to colonise the agricultural landscape by a new productive order of beings. This pre-scripted purpose was embedded in biotechnology's DNA, holding apart and intact the principle of descent from that of an organism's environmental interaction. Biotechnology came to do a certain job, in a context of mute farming environments, and unperceived human consumer publics.

In the neo-liberal context of World Trade Organisation parameters for food politics, environmental science was the only means to challenge what could otherwise be debated in terms of political economy of global food supply. The non-human thereby became analogically potent, carrying so many human stakes that were not legitimised in the division operated by scientific fact/cultural value. The argument presented in this chapter is that when the discourse of scientific knowledge about the non-human world was made so paramount in the framing of the GM question, and public debate was sectioned off from contaminating the influence of scientific fact, it would seem that the 'naturalist' cosmology (Descola) was dominant. However, it appears that this cosmology cannot accommodate the public's associative instincts, and analogical projections of common interests and 'kinship' (in non-genealogical manner) with non-human co-residents and protagonists rose from their slumbers as possibilities for non-human relationality.[19]

This chapter's approach has been to take a post-exceptional view of kinship, that looks at the implications of removing barriers to what can be thought of as 'natural relation'. In the case of GM foods this involves a radical interjection of choice and the making of optative relationship with what was merely taken for granted as the countryside – a place where city people could 'get away from it all', and imagine themselves, their 'selves', in surroundings where life processes and tenacious species have found refuge from the advance of modernity in hedgerows, abandoned railway-lines, and agriculturally unprofitable swamps and moors. In congruence with processes of subjective individuation and 'self-help' personhood, kinship as developed through comparative anthropology in terms of directly faced and compelling contexts of solidarity, exchange and desire, can come to the aid of imagining new social relations to the non-human. Post-exceptional kinships can be found 'in the making' by attending to the fertile conjunctures of relational selves and landscapes of animated presences, that are made visible in the new terrain of the politics of natural difference. If the building blocks of life can be reorganised at will, and the harvesting of energy from primary biotic production has become possible to re-engineer through selective implanting of genetic resistance to herbicide, this has simultaneously brought about the conditions for displacing the notion of kinship from its identity status of being a purely human concern.

Notes

[1] See Whitehouse this volume for a discussion of the key notion of 'balance' in reconciling alternative modes of valuing non-human species.

[2] From http://www.aebc.gov.uk/aebc/reports/reports.shtml accessed 8th March 2004.

[3] I thank my colleague at Durham, Thomas Widlock, for the observation that this kind of debate in Germany would have been expected to have been conducted in

parliament. Key among the advocates of this public debate was the Agriculture Minister, Michael Meacher, from the supposedly 'old left' wing of the labour Party.
[4] There is not room in this chapter to go into the topic of transgenic pollination: whereby the genes of modified crops such as oil seed rape might seek out relatives of the *brassica* genus in the British countryside, and could be anticipated to spawn 'superweeds'.
[5] All the major circulation national newspapers, and my regional paper the Yorkshire Post, were covered in this survey. Additionally, the Farmers Weekly, Farmers Guardian and New Scientist were included.
[6] Telegraph 17.10.03 p.3.
[7] Guardian 17.10.03
[8] Independent 17.10.03 p.20.
[9] The Times 17.10.03 p.9.
[10] Daily Mail 17.10.03 p.6-7.
[11] Independent 17.10.03 p.20.
[12] (Independent 17.10.03 p.5).
[13] Independent 17.10.03 p.20.
[14] New Scientist18.10.03 p.8
[15] Daily Telegraph 17.10.03 p.27
[16] Mohr reports more favourably on the experience from Denmark: "political participation in Danish consensus conferences is relatively high and politicians have a clear understanding of the role they play in communicating the product of the consensus conference network to the political arena." (Mohr 2002: 161)
[17] Guardian 20.02.04 p.8.
[18] It is not necessarily about assigning interiority to non-humans, but extending the logic from unreflexive naturalism to recognise among other things the dependence of moderns on their experiences of interiority generated by contact with nature interaction (Descola 2005: 336, Mabey 2005). See Coetzee, 1999 and Albie 2004 for fictional and dramatic exploration of these themes.
[19] For a view of how science also exploits sentiment towards animals, see Candea's work on meerkat research (Candea 2010.)

Bibliography

AEBC 2001 "Crops on Trial", Agriculture and Environment Biotechnology Commission.

Albie, E. 2004 *The Goat or "Who Is Sylvia?"*. London: Methuen.

Argyrou, V. 2005 *The Logic of Environmentalism: Anthropology, Ecology and Postcoloniality*. New York: Berghahn.

Bauer, M. & G. Gaskell (eds.) 2002 *Biotechnology: the Making of a Global Controversy*. Cambridge: Cambridge University Press.

Bird-David, N. 1993 "Tribal Metaphorisation of Human-Nature Relatedness: A Comparative Analysis". In K. Milton (ed.) *Environmentalism.* London: Routledge: 111-124.

Campbell, B. 2005 "On 'Loving Your Water-Buffalo More than Your Own Mother': Relationships of Animal and Human Care in Nepal". In J. Knight (ed.) *Animals in Person.* Oxford: Berg: 79-100.

—. 2008. "Fields of Posthuman Kinship". In J. Edwards and C. Salazar (eds.), *European Cultures of Kinship in the Age of Biotechnology.* Oxford: Berghahn: 162-178.

Candea, M. 2010 "'I Fell in Love with Carlos the Meerkat': Engagement and Detachment in Human-Animal Relations". *American Ethnologist.* 37 (2): 241-258.

Coetzee, J. 1999. *The Lives of Animals.* Princeton: Princeton University Press.

Cronon, W. (ed.) 1995. *Uncommon Ground: Toward Reinventing Nature.* New York: Norton & Co.

DEFRA 2003 "Modelling the Effects on Farmland Food Webs of Herbicide and Insecticide Management in the Agricultural Ecosystem". Research Report EPG 1/5/188.
http://www.defra.gov.uk/environment/gm/research/epg-1-5-188.htm

Descola, P. 2005 *Par-delà Nature et Culture.* Paris: Editions Gallimard.

—.1996 "Constructing Natures: Symbolic Ecology and Social Practice", in P. Descola and G. Pálsson (eds.), *Nature and Society: Anthropological Perspectives*, London: Routledge: 82-102.

Doniger, W. 1999 'Reflections'. In J. Coetzee. *The Lives of Animals.* Princeton: Princeton University Press: 93-106.

Escobar, A. 1999 "After Nature: Steps to an Anti-Essentialist Political Ecology". *Current Anthropology* 40(1): 1-30

ESRC Global Environmental Change Programme 1999 *The Politics of GM Food: Risk, Science and Public Trust*, Special Briefing No.5, University of Sussex.

Evans-Pritchard. E. 1950 "Kinship and the Community among the Nuer" In A. Radcliffe-Brown and D. Forde (eds.) *African Systems of Kinship and Marriage.* Oxford: Oxford University Press: 360-391.

Feeley-Harnik. G. 2001 "The Ethnography of Creation: Lewis Henry Morgan and the American Beaver" In S. Franklin & S. McKinnon (eds.), *Relative Value: Reconfiguring Kinship Studies.* Durham, Duke University Press: 54-84.

—. 2007 "An Experiment on a Gigantic Scale: Darwin and the Domestication of Pigeons". In R. Cassidy and M. Mullin (eds.) *Where the Wild Things are Now*, Oxford, Berg: 147-182.

Franklin, S. 2001 “Biologization Revisited: Kinship Theory in the Context of the New Biologies”. In S. Franklin and S. McKinnon (eds.) *Relative Values: Reconfiguring Kinship Studies.* Durham, Duke University Press: 302-327.

Franklin, S and S. McKinnon (eds.) *Relative Value: Reconfiguring Kinship Studies.* Durham, Duke University Press.

Gottweis, H. 1998 *Governing Molecules. The Discursive Politics of Genetic Engineering in Europe and the United States.* Cambridge, MA: MIT Press.

Haraway, D. 2008. *When Species Meet.* Minneapolis: University of Minnesota Press.

—. 1997 *Modest_Witness@Second_Millennium. FemaleMan_Meets_ OncoMouse.* London: Routledge.

Jones, S. 2000. *Almost Like a Whale: The Origin of Species Updated.* London: Anchor.

Klüver, L. 1995 “Consensus Conferences at the Danish Board of Technology”. In S. Joss and J. Durant (eds.) *Public Participation in Science: the Role of Consensus Conferences in Europe..* London, Science Museum: 41-49.

Knight, J. (ed.) 2005 *Animals in Person: Cultural Perspectives on Human-Animal Intimacies.* Oxford: Berg.

Latour, B. 2004 *The Politics of Nature: How to Bring the Sciences into Democracy.* Cambridge: Harvard University Press

Lévi-Strauss, C. 1973 [1962]. *Totemism.* Harmondsworth: Penguin.

Mabey, R. 2005 *Nature Cure.* London: Pimlico.

Mayer, I. 1997 *Debating Technologies. A Methodological Contribution to the Design and Evaluation of Participatory Policy Analysis.* Tilburg: Tilburg University Press.

Milton, K. 2002 *Loving Nature: Towards an Ecology of Emotion.* London: Routledge.

Mohr, A. 2002 “A New Policy-Making Instrument? The First Australian Consensus Conference”. Thesis. Griffith University

Novellino, D. 2007. “Talking about ‘Cultura’ and ‘Signing Contracts’: The Bureaucratization of the Environment on Palawan Island (the Philippines)”. In C. Maida (ed.) *Sustainability and Communities of Place.* London and New York: Berghahn: 82-107.

Nowotny, H., P. Scott, and M. Gibbons 2001 *Re-Thinking Science: Knowledge and the Public in an Age of Uncertainty.* London: Polity Press.

Olwig, K 1993 “Sexual Cosmology: Nation and Landscape at the Conceptual Interstices of Nature and Culture. Or What Does

Landscape Really Mean?" In B. Bender (ed.) *Landscape: Politics and Perspectives.* Oxford: Berg: 307-343.

Radcliffe-Brown, A.R. 1951 "The Comparative Method in Social Anthropology". *Journal of the Royal Anthropological Institute.* 1(2):15-22.

Rival, L. (ed.) 1994 *The Social Life of Trees: Anthropological Perspectives on Tree Symbolism*, Oxford: Berg.

Smuts, B. 1999 "Reflections". In J. Coetzee. *The Lives of Animals.* Princeton University Press: 107-120.

Weston, K. 2001 "Kinship, Controversy, and the Sharing of Substance: the Race/Class Politics of Blood Transfusion". In S. Franklin and S. McKinnon (eds.), *Relative Values: Reconfiguring Kinship Studies.* Durham: Duke University Press: 147-174.

Willis, R. 1994 *Signifying Animals: Human Meaning in the Natural World.* London Routledge.

CHAPTER NINE

OLD MACDONALD HAD A PHARM: ANIMALS, SCIENCE AND PHARMACEUTICAL PRODUCTION

CATHRINE DEGNEN

Introduction

This chapter examines the juncture of animals and science through the lens of pharming. "Pharming" is a blanket term that covers a wide range of techniques, but broadly speaking it entails the use of genetic engineering to produce pharmaceuticals for human consumption in non-human organisms. Pharming consists of inserting genetic material from one species into the genome of another in order to produce desired characteristics, and thus creating transgenic animals and plants.

My particular interest here is in pharm *animals* and the various relationships that pharming evokes. As pharming involves the crossing of species boundaries, one obvious relationship is that of transgression: a category breach that prompts worries over the boundaries between nature and culture writ large. While animals, and especially farm animals, have long been put to work by humans as a supposedly natural resource to be exploited, the genetic engineering of plants and animals arguably moves into new territory that prompts worries over acceptable limits and responsible social action. This is the dilemma so often discussed in the social science literature and amongst members of the public as they both grapple with the social and cultural implications of transgenic biotechnologies.

A second relationship is less obvious because it is generally left unspoken. It is however one I wish to call attention to in this chapter. This is the layers of cultural significance that the term pharming carries with it by virtue of its close affiliation with "farming". On the one hand, this relationship is simply an innocent pun, a verbal joke, which plays on the similarity in spoken English between "ph" and "f" to incorporate "pharmaceuticals" into "farming". On the other hand, farm animals and farming occupy a very particular place in Western cultural beliefs, and so by virtue of being linked to farming, pharming thus inescapably shares and carries some of these meanings, too. That is to say, "pharming" is not a neutral term, but one already laden with significance.

Finally, a third relationship that pharming evokes has to do with public perceptions of norms of comportment for genetic scientists, a theme presaged by Stapledon's *Sirius* as described in the Introduction to this volume. When considering the implications of genetic engineering notions of integrity, knowledge, and acceptable human behaviour in science were at stake for the people who participated in the larger research project from which this chapter draws.Pharming thus brings front and centre the human relationships at the heart of scientific endeavour as well as the relationship between science and observing publics.

In order to contextualise these three relationships, I begin by briefly outlining the recent history of pharming. I consider some of the cultural boundary issues that pharming raises, such as the simultaneous alikeness *and* difference of animals and humans. Humans and animals must be perceived as sufficiently alike in order for transgenic technology to be efficacious and worth pursuing in the laboratory. At the same time however, animals must be somehow held sufficiently apart and subservient to humans in order that they may be used in the pursuit of human benefit. I then look at the intersection of biowealth, patenting and gender in one example of the naming practices of some pharm animals. Finally, I query the puzzling absence of pharming in recent British debates over another kind of genetic engineering, that of GM plants, food, and crops, and turn to data from ethnographic fieldwork I conducted on public understandings of genetic modification in order to more fully explore some of these tensions.

Pharming

Unlike the genetic modification of food, pharming has not been a topic of widespread discussion or debate in Britain. It was however during ethnographic fieldwork on public understandings of genetically modified food in 2003-2004 that I first came across the term when I conducted a literature review on genetic engineering and food. Genetic engineering is being tested and used on animals for a variety of purposes: as disease models for use in medical and biological research; as a source of organs and tissues for human transplants such as experiments carried out on pigs

(Radford 2000); for novel pets such as the GloFish[1] or to clone existing pets (such as the cloning service offered by "Genetic Savings and Clone" in the United States); to improve productivity in agriculture by making animals with higher rates of milk, meat and wool production; to change characteristics like of milk for people who are lactose intolerant or to heighten the tolerance of farmed salmon to colder conditions; and as so-called "bioreactors" for drug production (Einsiedel 2005; GeneWatch 2002: 6, 49, 55; Niemann *et al.* 2005; Urbanik 2007: 1207-8).

It is the last of these forms of pharming that is my main focus here: the use of genetic engineering in animals to produce drugs for human consumption. Niemann *et al.* clarify that pharming "entails the production of recombinant biologically active human proteins in the mammary glands of transgenic animals…The mammary gland is the preferred production site, mainly because of the quantities of protein that can be produced in this organ" (2005: 286). This point is reinforced by Bauman *et al.* who detail the "incredible level of organisation" in mammary glands with their "remarkable ability to convert circulating nutrients into milk components" (2006: 1235); "the lactating mammary gland", they write, "is a prodigious protein-producing factory. Therefore, it is little wonder that with the advent of genetic engineering of mammals someone would propose making use of the mammary gland to produce large quantities of proteins not normally found in milk" (*Ibid*: 1241). Mammaries and milk are repeating themes in pharming, and ones I will return to later on in this chapter.

During the initial stages of writing this piece in 2005, unlike genetically modified plants, crops and micro-organisms, all genetically modified animals were in developmental stages and none were in commercial use (GeneWatch[2] 2002:58). However, in June 2006 the European Medicines Agency (EMEA) reversed its initial ruling against licensing a drug called Atryn and gave the go-ahead for commercial use of the drug in Europe, pending final approval (BBC 2006). Atryn, an anti-clotting agent, is being produced by a Boston based company called GTC Biotherapeutics by genetically engineering goats so that the agent is made in their milk[3].

Many different kinds of animals have been used in the development of pharming techniques and products. In addition to goats, these include sheep, mice, pigs, hens, and cows. Research experiments on transgenesis in animals dates back to the 1970s, with a mouse as the first transgenic animal to be produced reported in the literature in 1980 (Kelly 2005:61). Pharming got off to a start in 1987 when researchers at the National Institute of Health in the United States "first produced a foreign protein in the milk of a transgenic animal: the human tissue plasminogen activator in mouse milk. Since then, at least 29 human therapeutic proteins have been produced in transgenic animals…, most of them in milk, but some in blood, urine, and sperm" (GeneWatch 2002:39).

These human proteins and host animals include sheep that have been used to produce a protein called alpha 1-antitrypsin in their milk (a protein needed for humans suffering from cystic fibrosis and emphysema); human proteins that seal wounds and stop bleeding; and a human enzyme that breaks down fats for people with pancreatic damage (Radford 2000). Transgenic mice have been produced with human proteins in their milk (Meek 2000), hens have been used to produce anti-cancer proteins in their egg whites (Radford 2000b), and cows developed with a protein in their milk to treat multiple sclerosis in humans (Clark 2000)[4].

Given their novel status and the ways in which their origins confound the English language–are they best described as bred, made, developed, grown, produced, or raised?–it can be difficult to know how to describe the genesis of pharm animals. The impetus behind some of these animals however is British, most notably by a firm called PPL Therapeutics. This was a commercial business based on the Roslin Institute site in Scotland, the same Roslin Institute linked famously to Ian Wilmut and Dolly the Sheep (Radford 2000)[5]. Other pharm animals have been developed in the United States by firms such as Genzyme and GeneWorks, in the Netherlands by a firm called Pharming, and in New Zealand by AgResearch. Countries that have permitted experimental research in pharming include Canada, China, Cuba, Finland, France, Germany, Hungary, Sweden, and Switzerland (Clark 2002; Radford 2000b; Taussig 2004:313; GeneWatch 2002:41-44). Pharming then is not insignificant in its global scale and reach.

To give some idea of scales of production and use, in the UK, scientific procedures on genetically modified animals grew by more than 800% between 1991 and 2000, from 62,445 to 581,740 (GeneWatch 2002:11). More than 98% of these procedures were on genetically modified mice but procedures on other animals species, not including mice, also increased over one hundred-fold from 639 to 6,580 (*Ibid*). While overall the numbers of animals used in research between 1970 and 2003 were in decline, the percentage of animals used in genetic engineering research is increasing according to figures reported in 2004 (Amos 2004). Furthermore, it was reported that for these figures relating to genetic modification,

> …the recorded procedures also count the breeding events that maintain the numbers of animals for labs to work on - not merely the scientific tests performed on those altered animals...The stats also take no account of

> "wasted" animals - animals bred for their tissues and then discarded or animals rejected because their genetic modifications did not work. If these were included in the annual statistics, the figures for animal use would be considerably higher (Amos 2004).

This is not dissimilar to the high rates of failure in attempts to clone animals; on average it takes about 150 to 200 failed attempts to create one successful clone (Morelle 2007).

Boundaries

The ways in which new genetic technologies present challenges to boundaries between nature and culture, as well as the nature of nature itself and the locally cultural ways in which biotechnology is digested, have sparked a burgeoning literature in anthropology (see for example Carsten 2004; Edwards *et al.* 1999; Edwards 2000; Franklin 1997, 2003a, 2007; Haraway 1997; Rabinow 1992; Rapp, 1999; Strathern 1992). The majority of attention to date however on this topic has been on new reproductive technologies for humans. My argument here takes inspiration from the theoretical framings of this literature, especially in regards to the ways in which the new genetics reorder old assumptions about the social and the biological, and about relatedness. I am particularly interested in how pharming reorders relationships between humans and animals via biotechnology. That is to say that while animals, and especially farm animals, have been long put to work by humans as a "natural resource" to be exploited, I propose that pharming arguably raises new questions about human-animal relations and different ways of knowing about these relationships.

In particular, it seems to me that pharming simultaneously raises the old issue of similitude/alikeness and difference/dissimilarity between animals and humans, but in novel forms. Imaginings of similarity between animals and humans have to persist in order for transgenic technology to be promoted and to be efficacious. Simultaneously however, animals have also to be sufficiently dissimilar from human beings in order to serve as human handmaidens and as living drug factories. The difference now is that these are conceptualised within genetic frameworks, but also as "bits and bytes of information" (Edwards 2002) that can be snipped here and inserted there, crossing species-specific boundaries, or, indeed, not.

There is of course no straightforward relation between animals and humans in Euro-Western societies. Humans are, within some frames of reckoning, one kind of animal. And yet, they commonly distinguish themselves from all other animals through evocation of higher thought processes, use of language, and the existence of culture. Contradictorily, animals are perceived in some contexts as lesser beings than humans, a natural resource to be exploited for human benefit, but in other contexts as full-fledged family members that are equal domestic residents in the family unit, such as pets.

Broad historic shifts have been posited in the understanding of relations between animals and humans. Tapper (1994), for example, describes a transformation in Western thought about animals from an anthropocentric view to a growing anthropomorphism. The anthropocentric perspective was current in the Middle Ages and the Renaissance. It considered that humans, made alone in the image of God and having the gift of reason, were entitled to use animals to serve their own purposes. The latter perspective, anthropomorphism, gained ground via the growth of natural history and the recognition of the physical parallels between humans and other animals, a perspective solidified and made real by Darwinian theory (Tapper 1994:47-48, drawing on Thomas 1983). Tapper argues that both positions are imperfect as they do not acknowledge either cultural [anthropomorphism] or species-specific [anthropocentric] biases (Ingold 1994b:9), and he uses the ethnographic record to demonstrate the impressive diversity in cultural worldviews on the matter.

I propose that conventional farm animals, with an "f" and not a "ph", in particular have an interesting place in this complicated mix. Perhaps like no other category of animal, they straddle multiple realms and embody the contradictions at heart of the boundary conundrums between categories of animal and human and the differing relations that mediate these categories (c.f. Leach 1964). On the one hand, farm animals are bred, raised, and reared for what they can contribute to human endeavour (a source of labour, food and materials), and in this sense are purely a source of profit (either economic or for human well-being) to be exploited in a highly unequal relationship. Indeed, within Western categorisations, it is human *ownership* (and the relationship of right of use that this connotes) that distinguishes farm animals from wild animals.

But on the other hand, consider the intriguing place farm animals occupy in the cultural imagination, particularly as linked to childhood. As the Department for Environment, Food and Rural Affairs (DEFRA) estimates that just under 500,000 people are employed by the farming industry (DEFRA 2008), only a very small percentage of the British population is engaged in any kind of farming practice. Indeed, the vast majority of Britons have little lived experience of contemporary farming life. Despite this, farms occupy a powerfully evocative position in both cultural beliefs and understandings of rurality and countryside as well as a remarkable role in Western childhood play. Sheep, geese, goats, horses, cows, pigs, chickens, farms, farm yards, barns, tractors and farmers form a vivid cast of characters populating cultural imaginings of the countryside as well as children's literature and children's toys. These draw however on highly romanticised and sanitized visions of the farm and countryside: notably absent are the abattoir, multinational conglomerates selling patented seeds and chemicals, the ageing demographics of farming communities, and bankrupt families who must auction off their livelihoods and family homes.

Farm animals and farm yards hold a prominent place in childhood as source material for songs, books, plastic toys and for teaching the mimicry of animal noises which parents encourage. As one example amongst many, Alton Towers, a major amusement park in the English Midlands, plays on these childhood visions of farm in advertising one of its attractions, Old MacDonald's Farm. It proudly states in its promotional material that:

> Animal lovers of all ages are welcome at Old MacDonald's, where farmhands are needed to get to know the different creatures, and bash some troublesome berries. This fantasy farmyard is the perfect place for kids to sing along with the animals, ride the tractors and take a boat ride, while learning all about the farmyard world (Alton Towers 2005).

I was unaware of Alton Towers' Old MacDonald's Farm when I chose the title for this chapter, but we are both evoking a children's song that recites the names of farmyard animals owned by Farmer MacDonald and that mimics the noises each animal makes. Like this song, some researchers have argued that editors of children's books often use animals to relate stories of social values and rules. This is, they propose, because children can often relate more easily to the feelings (real or imagined) of animals than they can do of other human beings (Serpell 1986 in Tapper 1994:50, see also Tapper 1994:51). Tapper suggests that "in a society which offers little experience of what animals are 'really' like, they become stylised or idealised humans" and thus the prominence of "pets, zoos, and animal toys, the prevalence of animals in children's stories, and the universal success of both animal cartoons and wildlife documentaries" (1994:56) become more explicable. Tapper further argues that in obligatory childhood visits to the zoo, what children and adults are most interested in is observing human-like behaviour and characteristics in animals such as "feeding, copulating, mothering, playing, and fighting", interests that are paralleled with stories about animals from "Beatrix Potter's *Peter Rabbit* to George Orwell's *Animal Farm*", none of which are necessarily about animals as animals, but about animals representing issues at stake in human relations and power struggles (Tapper 1994:57).

In contrast, farm animals on working farms are working animals, and are used by humans to promote human interests. However, the main point of contact the vast majority of British people have with farm animals is not on farms. Rather, it is as characters from childhood play in which animals are humanised, becoming sources of lessons on social values and morals. Such practices soften the boundary between "human" and "animal" and blur easy distinctions between categories.

The concept and practice of pharming in turn blurs this boundary one degree further by substituting animal bodies for human ones in order to grow substances such as human enzymes and proteins but not in human bodies. Pharming thus conceptually unites conventional and romanticised notions of farmyards and the familiar cast of characters inhabiting them in a play-on-words with unconventional and perhaps uncomfortable new kinds of husbandry and harvest. I propose that pharming can be used as a meeting point in which to examine the juncture of animals, science, and public understandings of genetics as well as the human activity and cultural imaginaries mediating these intersections. In the case of genetic modification, what particularly interest me here are arguments over norms, relationships, categories and forms of knowledge

Five Little Piggies: Naming, Gender, and Biowealth

One important series of relationships forged through pharming that I wish to examine more closely here are clearly evident in a news story covered in the British press. On March 15, 2000, the following news item was reported in *The Guardian*, a daily British newspaper:

> Five pigs born 10 days ago in Blacksburg, Virginia - but only announced yesterday by a British firm based at Roslin in Scotland - are clones that could lead to new ways to save life. Millie, Christa, Alexis, Carrel and Dotcom are living proof of the first step towards supplies of identical pig organs genetically engineered to be compatible with the human bodies that might need them. In a world where patients in urgent need of spare part surgery outnumber potential donors by 20 to 1, the five piglets and their successors are claimed to be "the only near-term solution to solving the worldwide organ shortage crisis"...The five pigs are proof that pigs can be cloned: they will enjoy a life of comfort and celebrity. Although born in the US, they were brainchildren of PPL Therapeutics, the commercial business built on the site of the Roslin Institute, progenitors of Dolly the sheep. Millie was the name long ago selected for the world's first pig clone.

> Christa honours Christian Barnard, who performed the first human heart transplant in 1967. Two others salute Alexis Carrel, a 1912 Nobel prizewinner who pioneered transplantation. Dotcom was so called for luck. "Any association with dotcoms right now seems to have a very positive influence on a company's valuation," said Ron James, PPL's chief executive (Radford 2000)[6].

This reportage immediately raises several points. The first is that despite claims the article makes to the assured celebrity status these piglets will accrue by virtue of being clones, several years on from their birth, they are to the best of my knowledge largely unknown to the British public. They are not revered nor evoked as iconic figures in the way that Dolly the Sheep is, despite her early demise. How and why particular figures in biotechnology remain present in everyday discourse and others fade into oblivion is a theme I will return to.

The second point I wish to highlight has to do with the naming of the animals – Millie, Christa, Alexis, Carrel and Dotcom – and the lessons that might be learned from these naming practices. It is, of course, quite usual for pets and for favourite farm animals to be named. Iconic animals in genetic science tend to be named as well. Dolly is an obvious example, but others that are again less well remembered include Tracy the sheep[7], developed to produce a protein (AAT) in her milk that would help treat lung diseases and Britney the hen, developing "anti-cancer drugs" in her eggs (BBC 2005).

But what of the sources of these names? I do not have information on either Tracy or Britney, but Jeanette Edwards has written how Ian Wilmut named Dolly "after the singer and song-writer Dolly Parton because the cell used to create her came from [what was quoted in *The Guardian* as] 'the impressive mammaries of another sheep'" (Edwards 1999:303). Strong traces of gender issues, unequal power relations, and casual misogyny seem apparent in this throw away comment from Wilmut. As for the five piglets, the genesis of their names was widely reported as the reasons behind them had been proudly detailed in the press release announcing their births. Millie, the only name recognisably female that is allowed to stay female, is explained away with long being the choice for the first pig clone of the Millennium, but all the other names, save the last, are identified as feminized versions of male scientists, thus embodying in their very names the men that pioneered the transplantation techniques they have been developed to extend, evoking a genealogy of biotechnological technique and knowledge through naming practices[8].

Sarah Franklin has written about the intersection of intellectual property rights, paternity and genetic capital in a recent work on kinship in the era of cloning (2007, 2003b). She uses Dolly the sheep to think through the myriad and fascinating implications of genetic engineering for genealogy and for gender. Particularly relevant from her work to my example of the piglets above is that Dolly was cloned using the technique of nuclear transfer, and that it is this technique that is patented. This technique permitted Dolly to be made from

> …the nucleus of a mammary cell, amplified from a frozen tissue sample taken from a pregnant Finn-Dorset ewe who had died six years earlier. This nucleus was inserted into an enucleated donor egg cell from a Scottish Blackface sheep. The resulting embryo was carried by two more sheep, the second of which gave birth to Dolly (Franklin 2003b:100).

Dolly thus disrupts traditional notions of genealogy, descent, and reproduction. She also "is the product of a fertile union among several females" which prompts the observation that perhaps "maternity has triumphed over paternity in a kind of recapitulation of the ancient matriarchy theories so influential in early feminism" (*Ibid*).

However, as Franklin further elaborates, Dolly is valuable only as a proof in flesh and blood of the successfulness of Wilmut's technique, a technique that can be patented. Thus, "the viability of the means of reproduction used to make her, nuclear transfer technology, is the source of new genetic capital… (and consequently, in) this sense, cloning by nuclear transfer enables genetic capital to be made alienable from the animal herself" (*Ibid*: 99). It is not what *is* (Dolly) that is valuable or patentable, but rather what Dolly demonstrates it is *possible to do* (nuclear transfer) that holds value. This in turn is why Franklin concludes that maternity does not triumph via nuclear transfer; it is instead, she concludes, erased and subordinated. In the age of patentable biotechnique, what becomes more important is the "scientific *paternity*…which lies with Wilmut and his colleagues…who designed the blueprint of the technique that made her a viable offspring to begin with" (*Ibid*: 101, emphasis added). Paternity "has not so much been displaced as dispersed into acts of scientific creation and principles of legal ownership" (*Ibid*: 100), producing in the words of the journalist above, patentable "brainchildren" of the male scientists who brought them into being in the first place. Gendered idioms of proprietorship and control are thus evoked through the naming practices of these pharm animals and the playings upon gender, genealogy and kinship evoked by cloning itself whereby "the gender of the new genetic capital is very familiarly paternal" (*Ibid*: 104).

The third point to make is why patenting is so relevant. This comes into focus in light of the fifth piglet's name, Dotcom, and the soaring corporate hopes of money to be made via pharming and cloning. Indeed, GeneWatch reports that "the annual world market for just one potential transgenic product - blood coagulating factor VIII - is estimated at $880 million, which could theoretically be produced by just one transgenic cow" (GeneWatch 2002:7). In a different but related realm of activity, "Genetic Savings and Clone", the America pet cloning company, seems to knowingly play on the "bankability" and potential profit of biotechnology in choosing its company name. The economic stakes of pharming and cloning are high.

In Franklin's work that I cite above, she has also written about PPL Therapeutics and how the technique of nuclear transfer "is a device for seeding a corporate plan for the production of biowealth in the form of what PPL Therapeutics describes as "bioreactors": in this case, the sheep that will function as living pharmaceutical producers by excreting valuable proteins in their milk" (Franklin 2003b: 101). Cloning techniques, such as those developed via Dolly the sheep, have also been envisaged as permitting the rapid production of a number of transgenic animals for pharming[9]. The genetic capital of animals has long antecedents, as Harriet Ritvo's work on animal breeders in 18th century Britain has shown (1995), but the new potential commodity value of pharming techniques is breathtaking in its scale. A GeneWatch report from 2002 highlights this potential biowealth: in 2002, "the annual market for just six of the products currently produced in transgenic animal milk is $2.6 billion…and supplying five of the six drugs listed - with a total market of $1.4 billion – would theoretically only require 24 cows or 105 goats" (GeneWatch 2002:39). In an article from 2005 published by a senior scientist from the Institute of Animal Breeding, part of the Federal Agricultural Research Centre of Germany, the projected value on the global market for pharmed proteins is claimed to exceed $1 billon US by 2008 and to reach $ 18.6 billon US by 2013. This brings to mind Michael Fischer's blunt assessment of one of the driving forces behind biotechnology. It is hype, he says, which "is essential to getting venture capital and consumer demand, without which the creation of speculative new technologies would not occur" (2003: 1). Whether these estimates of global value have succumbed to hype, or whether they are based on modest and supportable figures, the amounts they suggest are truly the stuff of corporate profit-making dreams, and go a long way towards explaining the amount of investment in pharming research and development despite its low success rates to date.

A final point to make from this material returns us to milk and mammary glands. In amongst all the rhetoric surrounding the vast potential wealth to be generated by pharming, I am struck by just how omnipotent milk is in human and non-human animal relations (see also Bunn, Dransart, Grasseni and Stephenson, all this volume). Pharming would seemingly be impossible without it. And yet, despite its overwhelming presence in and necessity for pharming, attention to the gendered aspect of mammary glands and milk production is absent. This erasure is a direct parallel with what Franklin describes for maternity and Dolly whereby in a neat trick what it *is* (milk and mammary glands, producible only by female animals) is rendered unimportant in comparison with what it *allows* (potential generation of vast wealth via new techniques). Thus, rather than building up respect for the maternal power of milk, maternity and the gendered character of *what is* is stripped away. It becomes instead displaced by the hype and speculation of what pharming might bring economically and to medical science, or the *what it allows*.

Ethnographic Reflections on Boundaries, Transgression, and Categories

As stated earlier, my interest in pharming stems from a recent period of ethnographic fieldwork. This research was carried out in the north of England and centred upon public understandings of genetically modified crops and food, but not genetically modified animals. During my fieldwork I worked with a large variety of people including a Young Farmers youth group, workers at a historical ornamental garden, workers and volunteers at a historical garden, allotment holders, food caterers, housewives, health food store owners, farmers, activists, and a wide range of professionals not involved in either food or science. Some of these people had no vested interest in genetic modification, and others (such as the activists, farmers, and health food store owners), were active in the debates over genetically modified food. The fieldwork was based in two areas: the Barnsley area in South Yorkshire, and the Wilmslow and Alderley Edge areas in Cheshire.

Although debates over the genetic modification of plants, crops and food were at the time circulating widely in the British press, few if any links were made in publications and the media between genetically modified crops and animals. I began to wonder why animals were not featuring in the debates over genetic modification in Britain[10]. I also wondered how the ethnographic data I was collecting on genetic modification might relate to the genetic modification of animals. Closer examination of my data revealed that while my research focused on public understandings of genetically modified food and crops, the intersection of animals and genetic modification did at times appear in various guises throughout my work. The organic dairy farmer I got to know was perhaps the most obvious example–the genetically modified-free status of his herd and the milk they produce are central to his marketing. I particularly like the cartoon on his delivery van that features the image of a sentient cow with a speech bubble that states “I’m a smart cow!” since she is GM free. Discussions with the family running this farm made clear that their major concern with genetic modification is that the feed they are giving their cows is organic and that it should not become polluted with genetically modified soya or maize. Genetic modification thus only threatens the cows (and *their* form of biowealth which is their organic status) via what they ingest rather than their genetic code being altered by gene splicing with other animals. During the summer of 2003 when a government sponsored public debate meant that the topic of genetically modified crops was constantly in the media, our conversation circled around the dropping prices milk was receiving nationally, lack of legislation to deal with contamination should a neighbouring farmer decide to raise a genetically modified crop, and the impact that would have on this organic farm’s business.

When I asked another farmer in the Barnsley area if he would call himself a dairy farmer, he said that he and his wife are “specialist milk producers slash consultants slash in the tourism business”. He proclaimed that “the British public is being totally conned because we are all eating genetically modified produce all the time”. He explained that British farmers are being told not to feed their cows genetically modified fodder, although dairy farmers do feed their cows soya beans and gluten. Since insufficient protein is grown for fodder in the UK, 70% of the fodder is imported from the US where genetic modification dominates. Furthermore, he pointed out, soya beans are also a major source of protein for pigs and they along with cows have been eating genetically modified soya for some time now. Unlike the organic dairy farmer, this man was largely pro-genetic modification and thinks that it is “in wide use already” anyway.

The Erasure of Genetically Modified Animals from Genetically Modified Food

Apart from these farmers and others that I worked with whereby discussions of genetic modification and animals arose in terms of the animals’ own food consumption, the topic of genetic modification and animals came up otherwise only in terms of Dolly the sheep and cloning. During the twelve months that I conducted fieldwork, however, while the media was saturated with genetic modification stories–*The Guardian* alone had 176 items on genetic modification during my fieldwork period– and a three-strand consultation process sponsored by the British government was underway. In the same time period, *The Guardian* had only two articles on pharming and genetically modified plants, and none on pharming and animals although the broad topic of cloning received more coverage than this.

If the genetic modification of plants and crops was causing such a furore, why was there not a similar outcry about genetically modified animals? In an attempt to explain this puzzling absence, I returned to key documents that were produced before, during, and after the debate, as well as some that were constitutive *of* the debate on genetic modification. This included the Nuffield Council on Bioethics report on GM crops (1999), the GM Science review panel reports (Department of Trade and Industry 2003b, 2004), the economic strand report by the Strategy Unity (2003), the report of the public debate (Department of Trade and Industry 2003a), and the "prompt" or "stimulus" materials provided as part of the public debate in order to "inform" the public (GM Nation? Steering Board 2003)[11].

All of these documents frame debate over genetic modification in terms of crops and plants. Animals are rarely mentioned, and if they are, the focus quickly returns to plants and crops. For example:

> 1.1 the new technologies usually called "genetic engineering" or "genetic modification" (GM) promise to revolutionise medicine, animal husbandry and agriculture. An optimistic view it that GM plants and foodstuffs will make great, possibly indispensable, contribution to reducing mass hunger. Yet the development of GM crops has recently caused widespread unease in the United Kingdom (UK) and other European countries… (Nuffield 1999:6).

Genetic modification from this point on in the report is taken to mean crops and plants, but with no explanation of why this choice was specifically made. Similarly, in the Strategy Unit's Economic review report on genetic modification, an introductory paragraph written by Margaret Beckett (the then Secretary of State for Environment, Food and Rural Affairs) reads:

> From food to clothing, from medicines to fuel, plants are essential for sustaining human life…A new technique to develop novel crops is genetic modification. Genetic modification has a range of applications for both plants and animals, but has also been central to many recent advances in

> biotechnology research. GM micro-organisms, for example, already produce a variety of vaccines, as well as insulin for the treatment of human diabetes...GM crops could offer significant benefits to both farmers and consumers, and even the environment...GM crops could also be used to produce pharmaceuticals or as a source of renewable energy (Strategy Unit 2003:3).

Although Beckett mentions animals, micro-organisms and genetically modified plant pharmaceuticals, pharm animals or genetically modified animals are not brought into the frame. Likewise, the following introduction written by the Strategy Unit mentions very briefly that in addition to genetically modified crops, other applications of genetic modification could possibly include:

> A wide range of non-food crops, which could include crops used as "factories" for a range of goods, such as the production of pharmaceuticals, industrial oils, or renewable materials; as well as crops which could be used directly in the production of energy or fuel (2003:7).

Again, although other kinds of genetic modification are mentioned here, and pharming in the sense of genetically modified plants is included, genetically modified *animals* are noticeably absent. The only other mention I could find of pharmaceuticals in the 116 page report was in a section explaining genetically modified crops in a wider biotechnological context which stated that "GM animals (especially GM mice) have been developed as models for humans in medical research" (2003:32) and that some examples of genetically modified organisms with "marketable characteristics" include "GM microbes, GM crops and GM animals that act as "factories", producing pharmaceutically-active proteins" (2003:32). No other attempt at including them (or excluding them explicitly) was made in these key documents framing the debate – instead, they became effectively non-issues by their absence.

While the reports above were perhaps targeted at a specialist audience, the materials for the public debate on genetic modification in Britain were written explicitly to help members of the general public engage with the topic and make up their own mind. From the very first page, the document explains that "This debate is the country's first ever, nationwide public discussion exploring issues around Genetic Modification (GM)" (GM Nation? 2003:1). Issues around genetic modification could be expected to cover not just plants and crops (and include animals), but scope for including animals gradually gets written out over the following few pages. However, like the other reports mentioned here, this is never enunciated as an intentional strategy. Genetically modified pharming is mentioned on page 15, but not in terms of animals:

> GM also makes it possible to produce pharmaceuticals more widely available to all by reducing the costs of producing them…Problems have already arisen where residues from GM "pharmaceutical" crops have contaminated food crops grown afterwards in the same field." (GM Nation? 2003:15; see also page 16).

One mention is made specifically of animals, but only in passing:

> …in addition to plants, animals are already being genetically modified and cloned for use in agriculture (GM Nation? Steering Board 2003:29).

Despite these few brief mentions, "pharming" is however listed in the glossary of the booklet, but only in terms of plants and defined as: "crops that are genetically modified for pharmaceutical purposes" (GM Nation? Steering Board 2003:41, emphasis added). Thus, the overall trend that emerges from these documents is that the issues to be considered in Britain about genetic modification centre around plants and crops. Occasionally pharming is mentioned, but again largely in relation to plants. Genetically modified animals are thus removed from consideration.

I believe that this framing could have had a snowball effect – once the topic of genetic modification was delineated as "being about" plants and crops, momentum took hold and shut down other avenues of inquiry that might instead have sought to widen out discussion to include genetically modified animals. Another contributing factor could be that genetically modified commercial crops were felt to be an immediate threat and had to be addressed right away whereas genetically modified animals were at the time only in development and not felt to be a pressing issue.

Challenging Science

Given the disappearance of transgenic animals from the British debate, why then was the response to a transgenic bull named Herman in a neighbouring European nation, the Netherlands, the source of public controversy? A recent article by Karen Sue Taussig explores how different layers of cultural knowledge and practice circulate in the ways in which genetics are understood in the Netherlands (2004). Working with both geneticists and anti-GM activists, she demonstrates how although genetic scientists definitively claim authoritative knowledge about genetic modification, "other groups and individuals continually engage in constructing competing and sometimes conflicting interpretations of the significance of genetics in contemporary Dutch society" (Taussig 2004:328). Herman "has a human gene inserted into his genome so that his female offspring will produce milk containing the human form of the protein lactoferrine… [which could be] used in medicines for fighting intestinal infections in human" (Taussig 2004:313), and came to public attention in the Netherlands in the early 1990s. He and the transgenic biotechnology he represents were the subject of a national campaign against genetic modification in animals. Taussig carefully traces the particularly "Dutch" contours of this controversy and demonstrates how the "category confusion" of genetically modified animals is a particular affront to Dutch values of "containing differences within known and accepted categories", a social structure which has aided the perpetuation of social tolerance as "an integral aspect of Dutchness, of Dutch national identity, and of the Dutch nation" (2004:320). Another crucial factor was the continuing importance of distinguishing Dutch from neighbouring Germans, and worries over parallels between Nazi science/eugenics and genetic modification, which are "deeply antithetical to the Dutch social ideal of tolerance" (2004:319). Finally, the place of cows as centrally iconic of the Dutch nation (2004:316) means that the anti-GM bull/steer campaign was able to play on notions of "biotechnology distort[ing] the integrity of cows…[and] also suggest that biotechnology distorts the integrity of Dutch national identity" (2004:317).

The comparison with Britain is instructive. Perhaps in Britain it was the Dolly factor, with her widely publicised genesis in a Scottish research institute, that meant the focus of attention linking animals and genetic engineering became placed onto cloning. While techniques permitting cloning and transgenesis are closely linked in the laboratory, this relationship did not hold the same weight in the imagination of British publics. In media reports, cloning was portrayed as a scientific "first", led by Britain, with much less emphasis on why the technique was being developed and its relationship with the pharmaceutical industry. Additionally, the furore in Britain over genetically modified crops and previous recent food scares might have contributed to shaping worries over genetic modification onto something people and animals might *eat* rather than of transgenic animals raised for harvesting drugs and organs *from*.

Unlike Taussig, I did not extend my research to genetic scientists. However, some themes from my fieldwork on public understandings of genetically modified food are certainly thought-provoking in relation to pharming. One of these is that the people who participated in my research articulated reflexive and highly critical perspectives on "Science" and "Scientists" writ large. The following examples from three separate interviews that I conducted with four different people ranging in age from their mid twenties to their mid fifties demonstrate this point[12]:

> Phoebe – Yeah. I would just rather see money spent looking into why people are so miserable and sick and mentally ill and what is it about our life styles that needs changing, and our kids who are becoming obese or bullying or anorexic, what's wrong in our schools and in our society? I'd rather see more *social* stuff and all this kind of laboratory, white coat, fiddling about with test-tubes, you know, big deal. Where has that got us? We're in a mess. Let's put some thinking into social stuff, really.
>
> -------------------------------------
>
> Nicola – They [scientists] get too excited in their own cleverness.
>
> -------------------------------------
>
> Jack – And breeding, that's where I think we can get what we need. Not through that type of thing, not that GM, breeding is safe, it's been done for years and hasn't hurt nobody, so why change it?
> Cathrine – Cause it has worked?
> Jack – And it's still working, you know, so why, and same with plants, that's exactly the same with cross and breeding, it's been done, but it's *safe*, it's not hurt nobody, has it? So why try doing things what we don't really understand? And we don't know what effect it's going to have in twenty, thirty years' time.
> Anthea - I think sometimes they do these things because they *can*. You know, science gets to that point and…I mean like cloning, you know, there really wasn't any…
> Jack – Benefit at all!
> Anthea – Well not really. They say, don't they, that all these cloned animals age, arthritis, they also said that there were an awful lot of these cloned animals born with defects and deficiencies that were destroyed and no-body ever knew about it, scientists knew that that had happened, you know, but they only ever told us about Dolly who was successful and the rest was very quiet And as to why you would ever want to do it with a human…
> Jack – Oh, no
> Anthea – …that I *really* can't get my head round at all, I really can't see why anybody would, I'm sure somebody will before very long but why they'd want to I really can't see.
> Cathrine – Apart from, as you say, because they can.
> Anthea – Yeah! And because somebody wants to take an experiment to its conclusion and prove that they can.

I am struck by the extent to which my interlocutors engaged in commentating both on the significance of the genetic technology itself ("what's *in* it?"; "why do we need it?") but also on the place of science more broadly: science and scientists as rushing, impatient and blind, as doing science for science's sake, to "prove that they can". From the descriptions of my research participants, "science" appears to be an unseeing, blind form of misplaced human endeavour. This is particularly the case when there are so many other pressing social ills that need remedying as Phoebe points out, and when we have processes that are historically proven to work well as Jack says. Another theme evident in Jack and Anthea's interview is the ways in which science is imagining to have dirty secrets and that there are things that "we" are and are not "told" about, particularly in regards to cloning of animals (but not pharming). Scientists are perceived here in a generalised way as an unsocialised human category of person, as not quite like us both in knowledge base but also in terms of basic notions of decency and goodness.

The following passages are excerpts from three different interviews that give the sense of reservation that many people expressed over scientific endeavour in genetic modification:

> Christine – But, I mean there's all the issues regarding allergies, isn't there and they're putting species into another and people not knowing where they're coming from and what they're eating is changed…
> Oliver – I mean also you, the way that they have to switch off the plant's own immune system in order to plant the gene in, they use, they were called vectors I think in one of the books I read but basically it's a virus.
> Cathrine – Oh yes, oh right.
> Oliver – They have to use a virus to actually *put* the gene in and there's all kinds of risks just associated with that little area of it.
> Christine – Well, our scientists don't understand viruses! It's, there's so many…really big questions that, it's, I, I can't get my head around how we're prepared to accept going ahead so quickly as it seems that we are with something that is so precarious!
>
> --------------------------------
>
> Jill - And these things like taking a fish gene and putting it into a lettuce to make it so that it's actually easier [to grow] or more profitable, I mean there's so much money involved in this and you get a bit dubious.
>
> --------------------------------
>
> Nicola – …I think the whole point is that they are experimenting and if you are experimenting (laughs) you're waiting for the outcome, aren't you?!

While some research participants felt that scientists and regulatory bodies are to be trusted because they are responsible, have the training, are the specialists and will monitor each other, not everyone agreed. At times, scientists were described as not necessarily knowing what they are doing themselves, something of that was a significant cause of concern for many research participants. Also of concern was the profit-motivation linked to genetic technologies, with misgivings being expressed about the commercial concerns and corporate interests behind the research. This calls to mind Fischer's point regarding the transformations in academia from twenty years ago when

> ...academic biologists steered clear of entanglements with corporate profit drives...(to now when) almost every successful academic biologist is involved in a company as a necessary means to protect patented discoveries and produce them in forms that are no longer merely experimental but can be used, licensed, traded, and put to therapeutic use (2007: 567).

The general public is savvy about these transformations in the ways in which science is practised, and is sceptical about the implications of them for their own futures.

Conclusions

Ironically, such disparaging remarks seem to place scientists in a different moral realm, one to which my research participants did not relate. Jeanette Edwards has written about how cloning serves as a "conceptual brake, a marker of limits" (1999:301) "to possibilities presented by scientific intervention in human reproduction" (1999:307). The reconfiguring of social relations and of kinship in the era of biotechnology have been central concerns in the literature in the new genetics. But what then are we to make of genetic technologies that are not geared towards human reproduction? Do they matter socially in the same way? What do *they* come to serve as markers of? While the lack of debate over transgenic animals in Britain provokes a host of questions, my work on public understandings of genetically modified food has highlighted public perceptions of scientists in biotechnology as operating in a different register than that of 'normal human' comportment. My co-conversationalists indicate discomfort with the boundary transgressions the scientists are perceived as making, and furthermore identify their actions as ones that negatively separate them out from the general public. Indeed, there is the underlying suggestion that perhaps these transgressions are even animalistic in the sense evoked by Midgley (1994). She describes a deep ambiguity in the notion of "animal", one that is inclusive of humans, pets, wildlife, farm animals and all the various relations they represent. The second use of "animal" is one that

> …excludes humanity – the notion of an animal stands for the unhuman, the anti-human. It is a symbol for the forces which we fear in our own nature, and do not regard as a true part of it. It displays those forces as continuous with ones which we fear in the world around us – with floods, earthquakes and volcanoes – and thereby dramatizes their power. By speaking of those forces as "animal", we imply that they are in some way alien to us, therefore incomprehensible. But the peculiar alarm which they produce suggests also that they are *not* altogether alien – that we too carry the seeds of them in our nature (1994:35)[13].

While her use of "our nature" is itself problematic, I think there is some food for thought here. This is because by the very act of embracing transgenic biotechnology, research scientists, governments and biotech companies are diagnosed by observing publics as endangering their own humanity. This is because they are perceived as dabbling in activities that are perhaps outside the bounds of humanity itself, and certainly as using techniques that push up hard and abrasively against these boundaries of and straining them to their limit (c.f. Wynne 1996). Pharming thus represents a dual transgression: firstly of the genetically modified plants and animals which not only transgress cross-species boundaries but indeed become troublingly quasi-human when used to produce human enzymes and proteins. Secondly, there is a generalised perception of scientists and other powerful actors as transgressing the boundaries of acceptable *human* behaviour by engaging in genetic modification and by getting "too excited in their own cleverness". Thus, not only are various species boundaries being transgressed via genetic modification, but so too are core social categories of integrity, knowledge, acceptability, and comportment, due to individual scientists' decisions to employ such practices. As such, scientists engaging in biotechnology run the risk of having not just their work come under scrutiny, but of also being themselves categorised as unhuman or anti-human by observing, and critical, publics.

Acknowledgments

I would like to thank Maggie Bolton and Penny Harvey for organising the very lively *Animals and Science* conference at Manchester University where this work first appeared, and to Maggie for her careful reading of and suggestions for this chapter. My thanks as well to conference participants for their helpful comments. The research presented here is part of a larger research project entitled "Public Understanding of Genetics: A Cross-Cultural and Ethnographic Study of the 'New Genetics' and 'Social Identity'". European Commission Fifth Framework Programme: Quality of Life and Management of Living Resources. Contract number: QLG7-CT-2001-01668.

Notes

[1] The GloFish is a genetically engineered zebrafish that glows fluorescently under UV light and is for sale in the United States (see http://www.glofish.com/), although it is banned for sale by the Department of Fish and Game in the state of California (Einsiedel 2005:153).

[2] GeneWatch, or GeneWatch UK, is a not-for-profit public interest group which aims to build public involvement in decisions about the use of genetic technologies and public understanding of the same technologies. It monitors developments in genetic science and their applications to food, animals, health and society, and publishes freely available regular briefing papers and reports detailing these developments. These documents provide a wealth of information not easily accessible elsewhere but the highly sceptical position of GeneWatch towards the application of biotechnology and towards the companies producing them must also be born in mind.

[3] As described on the New York Times Business (2008) webpage under "Company Information", "GTC Biotherapeutics, Inc. is engaged in the development and production of human therapeutic proteins through transgenic technology that enables animals to produce a recombinant form of a specified human protein in their milk. Using this technology, the Company is developing a portfolio of recombinant blood proteins to treat a range of genetic and acquired blood deficiencies, including haemophilia and other blood coagulation disorders. It has also initiated the development of a portfolio of monoclonal antibodies (MAb's), for use as potential follow-on biologics targeted at several market products. Its first product ATryn, its recombinant form of human antithrombin, validated its transgenic production technology's capability. In February 2008, it announced that ATryn had met the statistical requirements for the primary endpoint in its trial to support its filing of a Biologics License Application (BLA), for the use of ATryn in United States."

[4] For a comprehensive list of animals, traits, companies involved, and countries where pharming research is occurring, see GeneWatch 2002:13-14; 41-44; 51-54.

[5] Since first writing this, PPL Therapeutics Plc has closed (Aldhous and Coghlan 2006) with some of the patents and licensing on products and technologies once owned by PPL being bought by Pharming Group of the Netherlands (Stamp 2007). "PPL" itself stands for "Pharmaceutical Proteins Limited" (Franklin 2007: 21).

[6] In 1993 PPL Therapeutics Plc, of the Roslin site, launched a US subsidiary, also called PPL, in Blacksburg Virginia to commercialise various patent applications made by the Virginia Polytechnic Institute (Glover and Tascarella 2003). The University of Pittsburgh Medical Centre's corporate biotech company, Regenecor Holdings Inc, then bought the Blacksburg based PPL in 2003 (Ibid).

[7] I first learned about Tracy from the BBC Science website, but was unaware until reading Sarah Franklin's *Dolly Mixtures* (2007) that the origin of the experiment

that created Dolly was an attempt to improve on the technique and the efficiency of the technique by which Tracy, born in 1990, was created (Franklin 2007: 35).

[8] This calls to mind Sharon Traweek's riveting account of the naming practices of machines and research groups in physics (1992). Her analysis of these practices reveals both the power relations at play within the international scientific community and the tropes of heroism and immortality at work. My thanks to Maggie Bolton for bringing this to my attention.

[9] My thanks to Maggie Bolton for reminding me of this other dimension in which cloning and pharming techniques abut and reinforce one another.

[10] My thanks to Penny Harvey for first calling my attention to this absence.

[11] The Royal Society has produced a lengthy report specifically on animals entitled *The Use of Genetically Modified Animals* (2001) but it likewise does not make a link between animals and genetically modified plants and food.

[12] All names have been changed to protect the anonymity of research participants.

[13] Michael (2001) makes a similar argument in regards to pharming when he writes that pharming consolidates a utilitarian perspective on animals, consequently also removing the fluidity of symbolic interpretation of animals in Western society, in turn limiting the symbolic complexity animals usually hold and through/against which human identities are also forged. The ultimate consequence of this, he argues, is that pharming increases the knowability of animals by essentialising them but also increases the unknowablility of ourselves as human beings.

Bibliography

Aldhous, P. and Coghlan, A. 2006 "Dolly's Cloning Revolution Fails to Materialise". *New Scientist*, July 1, 2006.

Alton Towers Website 2005 http://www.alton-towers.co.uk/themepark /guide_oldmacdonald.asp. Viewed 05.06.13.

Amos, J. 2004 "Modified Animals Continue Trend" http://news.bbc.co.uk/1/hi/sci/tech/3633468.stm Sept 7, 2004. Viewed 07.02.16.

Bauman, D.E., I.H. Mather, R.J. Wall and A.L.Lock 2006 "Major Advances Associated with the Biosynthesis of Milk". *Journal of Dairy Science* 89: 1235-1243.

BBC 2006 "Go Ahead for 'Pharmed' Goat Drug". *BBC News Website*. June 2, 2006. http://news.bbc.co.uk/1/hi/sci/tech/5041298.stm Viewed 06.10.27.

—. 2005 http://www.bbc.co.uk/science/genes/gene_safari/pharm/a_phar ming.shtml Viewed 05.06.16.

Carsten, J. 2004 *After Kinship*. Cambridge: Cambridge University Press.

Clark, A. 2002 "Fresh Blow for 'Dolly' Firm". *The Guardian*. March 19, 2002. http://www.guardian.co.uk/business/story/0,,669924,00.html. Viewed 07.02.27.

—. 2000 "PPL Moves from Dolly to Cows". *The Guardian*. December 15, 2000. http://business.guardian.co.uk/print/0,,4105876-108725,00.html. Viewed 07.02.26.

DEFRA 2008 https://statistics.defra.gov.uk/esg/quick/agri.asp. Viewed 08.08.15.

Department of Trade and Industry 2003a *GM Nation? The Findings of the Public Debate*. Crown Copyright.

—. 2003b *GM Science Review First Report: An Open Review of the Science Relevant to GM Crops and Food Based on Interests and Concerns of the Public*. Crown Copyright.

—. 2004 *GM Science Review Second Report: An Open Review of the Science Relevant to GM Crops and Food Based on Interests and Concerns of the Public*. Crown Copyright.

Edwards, J. 2002 "Bits and Bytes of Information". In N. Rapport (ed.), *British Subjects: An Anthropology of Britain*. Oxford and New York: Berg, 2002:163-179.

—. 2000 *Born and Bred: Idioms of Kinship and New Reproductive Technologies in England.* Oxford: Oxford University Press.

—. 1999 "Why Dolly Matters: Kinship, Culture and Cloning". *Ethnos* 64(3):301-324.

Edwards, J., S. Franklin, E. Hirsch, F. Price, M. Strathern (eds.) 1999 *Technologies of Procreation: Kinship in the Age of Assisted Conception.* 2nd ed. London: Routledge.

Einsiedel, E.F. 2005 "Public Perceptions of Transgenic Animals". *Revue Scientifique et Technique – Office International des Epizooties.* 24(1):149-157.

Fischer, M.M.J. 2007 "Four genealogies for a Recombinant Anthropology of Science and technology", *Cultural Anthropology* 22(4): 539-615.

Fischer, M.M.J. 2003 *Emergent Forms of Life and The Anthropological Voice*. Durham: Duke University Press.

Franklin, S. 2007 *Dolly Mixtures: the Remaking of Genealogy*. Durham: Duke University Press.

—. 2003a "Re-thinking Nature-Culture". *Anthropological Theory* 3(1): 65-85.

—. 2003b "Kinship, Genes and Cloning: Life after Dolly". In A Goodman, D Heath and M Lindee (eds.) *Genetic Nature/Culture* . Berkeley: University of California Press: 95-110.

—. 1997 *Embodied Practice: A Cultural Account of Assisted Conception.* London: Routledge.

GeneWatchUK 2002 *Genetically Modified and Cloned Animals: All in a Good Cause?* Derby, UK: Interleith.

Glover, L. and P. Tascarella 2003 "UPMC Buying Biotech Firm PPL Therapeutics". Pittsburgh Business Times, April 9, 2003. http://www.bizjournals.com/pittsburgh/stories/2003/04/07/daily30.html. Viewed 08.09.10.

GM Nation? Steering Board 2003 *GM Nation? The Public Debate*. Crown Copyright.

Haraway, D. 1997 *Modest_Witness@Second_Millenium.FemaleMan©_Meets_Oncomouse*. London: Routledge.

Ingold, T. 1994 "Introduction" In T. Ingold (ed.), *What is an Animal?* London: Routledge.

Kelly, L. 2005 "The Safety Assessment of Foods from Transgenic and Cloned Animals using the Comparative Approach". *Revue Scientifique et Technique–Office International des Epizooties.* 24(1): 61-74.

Leach, E. 1964 "Anthropological Aspects of Language: Animal Categories and Verbal Abuse". In E. H. Lenneberg (ed.) *New Directions in the Study of Language*. Cambridge: MIT Press: 23-63.

Meek, J. 2000 "New Age of Cloning for Health Care". *The Guardian.* June 29, 2000. http://www.guardian.co.uk/uk_news/story/0,,337639,00.html. Viewed 07.02.27.

Michael, M. 2001 "Technoscientific Bespoking: Animals, Publics and the New Genetics". *New Genetics and Society* 20(3): 205-224.

Midgley, M. 1994 "Beasts, Brutes and Monsters". In T. Ingold (ed). *What is an Animal?* London: Routledge.

Morelle, R. 2007 "A Decade on from Dolly". BBC News, February 23, 2007. http://news.bbc.co.uk/1/hi/sci/tech/6359011.stm Viewed 07.02.27.

New York Times Business 2008 "Company Information: GTC Biotherapeutics, Inc". http://topics.nytimes.com/top/news/business/companies/gtc-biotherapeutics-inc/index.html. Viewed 08.09.08.

Niemann, H., W. Kues and J.W.Carnwath 2005 "Transgenic Farm Animals: Present and Future." *Revue Scientifique et Technique–Office International des Epizooties.* 24(1): 285-298.

Nuffield Council on Bioethics 1999 *Genetically Modified Crops: The Ethical and Social Issues.* London: NCoB.

Rabinow, P. 1992 "Artificiality and Enlightenment: From Sociobiology to Biosociality". In J. Crary and S. Kwinter (eds.), *Incorporations.* New York: Zone Books: 234-252.

Radford, T. 2000 "Pigs Cloned for Organs down at the 'Pharm'". *The Guardian.* March 15, 2000. http://www.guardian.co.uk/uk_news/story/0,,183332,00.html. viewed 07.02.26.

Radford, T. 2000b "GM Hen's Eggs put on Cancer Cure Trail". *The Guardian*. December 4, 2000. http://www.guardian.co.uk/uk_news/story/0,,406470,00.html. Viewed 07.02.26.

Rapp, R. 1999 *Testing Women, Testing the Fetus: the Social Impact of Amniocentesis in America*. New York: Routledge.

Ritvo, H. 1995 "Possessing Mother Nature: Genetic Capital in Eighteen-Century Britain". In J. Brewer and S. Staves (eds.), *Early Modern Conceptions of Property*. London: Routledge: 413-426.

The Royal Society 2001 *The Use of Genetically Modified Animals*. London.

Serpell, J. 1986. *In the Company of Animals*. Oxford: Basil Blackwell.

Stamp, G 2007 "The Appliance of Science into Profit". http://news.bbc.co.uk/1/hi/business/6210100.stm. Viewed 08.09.10.

Strategy Unit 2003 *Field Work: Weighing up the Costs and Benefits of GM Crops*. London:HMSO.

Strathern, M. 1992 After *Nature: English Kinship in the Late Twentieth Century.* Cambridge: Cambridge University Press.

Tapper, R. 1994 "Animality, Humanity, Morality, Society." In T. Ingold, (ed.), *What is an Animal?* London: Routledge: 47-62.

Taussig, K. 2004 "Bovine Abominations: Genetic Culture and Politics in the Netherlands". *Cultural Anthropology* 19(3):305-336.

Thomas, K. 1983 *Man and the Natural World: Changing Attitudes in England 1500-1800.* London: Allen Lane.

Traweek, S. 1992 "Border Crossings: Narrative Strategies in Science Studies and amongst Physicists in Tsukuba Science City, Japan". In A. Pickering (ed.), *Science as Practice and Culture*, A. Pickering, ed. Chicago: University of Chicago Press: 429-466.

Urbanik, J. 2007 "Locating the Transgenic Landscape: Animal Biotechnology and Politics of Place in Massachusetts". *Geoforum* 38:1205-1218.

Wynne, B. 1996 "May the Sheep Safely Graze?: A Reflexive View of the Expert–Lay Knowledge Divide". In S. Lash, B. Szerszynski, B. Wynne (eds.), *.Risk, Environment and Modernity: Towards a New Ecology* Sage: London: 44 – 83.

AFTERWORD

SARAH GREEN

The key contribution of this collection is that it critically explores the hierarchies between differing practices and ways of understanding the relations between humans and animals. In that context, science, and the techniques and technologies of science, are contrasted, compared, brought into relation and distinguished from a variety of other ways of generating knowledge, classifying the world and carrying out many different practices. In all the chapters, it is clear that scientific thinking has a high formal position in the contexts explored, even where the science is questioned. Some chapters explore how such hierarchies were established (Nimmo and Bolton); others explore the diversity of ways that these differences are expressed (Stephenson, Bunn, Dransart and Degnen); and others explore the contingencies of these differences and distinctions, focusing on moments when outcomes and truths of the debate about the relations between science, animals and people seemed somewhat less certain than usual (Grasseni, Campbell and Whitehouse).

Collectively, the chapters demonstrate that human-animal relations (or enmities) have been used for all kinds of political purposes, both in the past and the present; that science has been used as a means to intervene a great deal in defining the parameters of the difference between humans and animals, which added to earlier approaches that focused mostly on property and labour in defining those differences; and that in more recent years, the scientifically defined distinctions between animals, humans and their activities and relations are being increasingly questioned. The questions are not only about the validity of scientific claims to truth; more often in fact, they seem to concern deeply moral and social questions about the implications of applying a scientific approach to human-animal relations. This is hardly surprising; as both the editors and several of the authors have pointed out, human-animal relations, as well as the practices of science, have been used both symbolically and politically to assert a rich variety of often competing moral and social truths for a very long time. And there are many texts available now that debate such matters, as discussed in the Introduction to this volume; what the anthropological and historical approaches in this collection can offer to this ongoing debate is a nuanced understanding of both the epistemological and moral/ideological questions that are involved: how the distinctions and relations between animal and human are made and maintained in a variety of different contexts and across a range of periods, and what political and economic implications there are in seeing those distinctions from one perspective rather than another.

Here, an important implication that arises from the approach taken in this book is that no perspective (scientific or otherwise) exists alone and unchallenged in the world, however dominant or powerful that approach may seem to be; there always appear to be other ways of understanding, other ways of generating the realities by which people live. The chapters by Bunn, Bolton, Degnen and Campbell particularly draw this out, though this is present in all the chapters in one way or another. So, given this existence of multiple ways of knowing and experiencing the world, which are often hierarchically ordered in different ways according to political, social and/or moral perspective, an important question arises: how should researchers handle the relationship between these different, and often competing, forms of knowledge? Strathern points out in *Partial Connections* (2004) that diverse knowledge practices inevitably use, or more often create, their own distinctive yardsticks (or scales, in Strathern's terms) in generating knowledge. To put it another way: the realities that a certain approach describes (e.g. the distinction between nature and culture) are also *defined* by that approach. As Nimmo's chapter outlines, and as confirmed by other chapters in this volume that explore different approaches, the distinction would not actually exist without the existence of that particular classification system and scientific approach. This may sound self-evident (science produces its own version of what is true), but its implications are often ignored when comparing different ways of generating knowledge: it is not possible to compare an indigenous way of understanding the characteristics of a camelid and an anatomist's understanding of the same animal (if, indeed, it is possible to say it *is* the same animal) without applying some kind of comparative yardstick. The researcher has to decide which yardstick to use: the indigenous one, the anatomist's one or a different yardstick taken from somewhere else (e.g. anthropology)?

Of course, this is a familiar problem in anthropology in general: which framework of knowledge should one use to understand other ways of knowing? The other way of knowing or the anthropologist's way of knowing? In the case of the interrelations between animals, people and science, the deeply political, social and moral elements which inevitably become involved in the debates over different perspectives add to the complexity of the question, and the chapters in this volume demonstrate a variety of ways to tackle it. They show how looking at the relationship between animals, humans and scientific knowledge and techniques is both a moral story and an epistemological one: the moral story concerns what we think is right and wrong about that relationship, and what we think is of value. The epistemological one tries to tease out how the different knowledges came to be constituted as such.

Importantly, several of the chapters demonstrate how claims to having a scientific approach often evade this moral, social and political issue by asserting that science is objective, and therefore value-free (i.e. there is no yardstick defining the scientific reality: the claim is that the reality is simply there). If there is one thing that the chapters in this volume demonstrate it is that no science in practice is value-free; and this becomes particularly obvious through the way the chapters highlight the co-existence of differing and frequently critical perspectives. In addition, there were several chapters where the science seemed to go wrong (e.g. Bolton's chapter) or where many people believed it was highly likely to go wrong (e.g. Campbell's and Degnen's chapters). This has some parallels with Callon's discussion of 'economics in the wild' (Callon 2007): Callon suggests that economics works in one way in theory, and in a rather unpredictable and different way when practiced in the world. This is, he argues, partly because of the tendency for one economic approach to confront alternative perspectives, as well as the incapacity of any theory to encompass everything that it describes and defines. Things go wrong. Grasseni's chapter is a particularly good example of how the implementation of one approach rather than another has unpredictable outcomes and involves a wide range of factors, not the least of which is the social implications of deciding to farm in one way rather than another.

This raises another element that appears in most of the chapters: that the relations between animals and people are actually *relations* as much as they are separations, and this applies not only to the many examples of non-European understandings of the engagement between animals and people, but also to people in Europe (e.g. Campbell and Degnen). That there are no hard and fast distinctions between nature and culture, scientific and social knowledge, or animal and human (what Campbell describes as 'post-exceptionalism') should no longer come as a particular surprise; what is somewhat more intriguing about this volume is how these blurred distinctions carry over into the practices of scientists and policy makers, who explicitly claim to make very clear distinctions. An example was Bolton's account of the way scientists working on improving llama health in Bolivia entirely ignored the official policy that there should be a sharing of knowledge between indigenous peoples and the scientists: given the widespread assumption on the part of what Bolton calls city-dwellers that people in rural areas were lacking in appropriate knowledge, there was no reason to think any insights could be gained by learning of the rural people's perspective. This might appear to be an insistence on maintaining a separation between science and indigenous knowledge; looked at another way, it involves ignoring the rules of science that things should be investigated before conclusions are drawn about them, and that social attitudes should not affect scientific work.

The one article concerning animals defined as wild (the grouse described by Whitehouse) draws out a similar issue, but defined in a slightly different way. The article describes a conflict between the interests of grouse hunters, people who herd sheep, and conservationists who want to protect raptors (which eat grouse). The issue concerned diverse efforts to define what had caused a steep drop in grouse numbers. The issue is set up as a 'problem' for science to solve, and the article defines a range of perspectives – all of which claim to be 'scientific' in one way or another, and which are mutually contradictory. In the articles describing domesticated animals, the conflict is usually local knowledge against the scientists; in the case of the missing grouse, this was partly the case (the gamekeepers had a claim to indigenous knowledge, as it were); but the problem was also set up as more of a puzzle than a conflict: given that there is less of a close relationship with grouse than there is with sheep and camelids, 'science' is recruited to resolve the mystery. The different results of research tended to reflect the interests and perspectives of the organizations involved, perhaps unsurprisingly.

That raises an important conclusion that can be drawn from this collection: while there may be a difficulty in comparing different knowledge practices, this does not prevent the existence of many overlaps between one approach and another; given that different ideas co-exist, it is common for people hold a variety of perspectives simultaneously, and also to borrow ideas from perspectives that are familiar. Campbell discusses the borrowing of Disney-esque notions from people who are concerned about the implications of introducing genetically modified crops in Britain; Degnen describes familiar Frankenstein-style fears about scientists running amok with their ideas and letting things get out of (social) hand(s). In both cases, the ideas expressed are being used in different contexts, and with different results, than the originals, as it were; but the ideas were borrowed from what was available. So in many senses, these two chapters suggest that people draw heavily from their historical and cultural context:

the media as well as the people Degnen interviewed interpret the activities of scientists using frameworks of understanding that are to hand for them – the parables of Frankenstein; the anthropomorphism of Disney; the fears inherent in *1984*; the tensions in understanding the relations between people, animals, plants, places, money, politics, history and the two powerful techniques of the contemporary period, technology and science. Most people of course make sense of their world using whatever intellectual resources they have to hand, a process that Dumit calls 'objective self-fashioning' (Dumit 1997). What is more interesting this volume is that it highlights the absence of singular perspectives: science does not, in fact, have an unquestioned place; the distinction between nature and culture does not go unchallenged in everyday life; and the scientists do this kind of symbolic bricolage just as much as everyday folk.

In that sense, one can conclude that if humans do not emerge as being exceptional in these texts, then neither does science. It is one amongst many different perspectives, with which it shares many characteristics. This is a point that Latour has made in a different context; here, the embedding of scientific practice within social and political relations is emphasised much more than the idea of science being embedded within a network of interests and vantage points. Interestingly, both Campbell and Degnen draw out how, over the last couple of decades, many people have gained a clear idea of 'the public' as a powerful entity that can act as a moral compass against the idiocies of politicians, scientists, and the owners of large companies. Somehow, 'the public' can, at times, speak, though the question remains open concerning what constitutes 'the public' (other than anyone who is not a scientific expert); the problem that remains, as with the operations of power in all contexts, is the possibility of being heard.

Bibliography

Callon, Michel. 2007. "What does it mean to say that economics is performative?" In D. A. MacKenzie, F. Muniesa, and L. Siu. (eds.), *Do economists make markets?: on the performativity of economics*. Princeton: Princeton University Press: 311-357.

Dumit, Joseph. 1997. "A Digital Image of the Category of the Person: PET Scanning and Objective Self-Fashioning." In G.L Downey and J. Dumit (eds.) *Cyborgs and Citadels: Anthropological Interventions in Emerging Sciences and Technologies*. Santa Fe, New Mexico: School of American Research Press: 83-102 ..

Strathern, Marilyn. 2004. *Partial connections, Updated Edition*. Walnut Creek, CA ; Oxford: AltaMira Press.

CONTRIBUTORS

Maggie Bolton came to anthropology after having studied physics and worked as an engineer. Her interests include the Andean region, science, colonialism and animal-human relations. She is based at the University of Aberdeen, and has published in journals such as *Journal of the Royal Anthropological Institute* and *The Sociological Review*. Her current research includes a research project on adult education and citizenship in Bolivia and she is also working on a monograph on llama herding and agricultural development.

Stephanie Bunn is Lecturer in Social Anthropology at the Department of Social Anthropology at the University of St Andrews, Scotland. She has carried out research among Kyrgyz pastoral nomads in Kyrgyzstan and Tajikistan over the past fifteen years. Her research includes nomadic relationships with the environment, perception and creativity, the notion of home, space, and textiles in the nomadic tent. Her most recent publication is *Nomadic Felts,* published by the British Museum Press 2010. She has also recently produced a web publication *Sound and Anthropology* (www.st-andrews.ac.uk/soundanth/) and translated and edited a volume on the research of Russian-Kyrgyz ethnographer, Klavdya Antipina.

Ben Campbell Ben Campbell is Lecturer in Anthropology at Durham University. He has worked since 1989 with Tamang-speaking communities of Nepal. He has published on exchange labour in village subsistence, environmental knowledge, politics of sustainability, and pilgrimage. At Manchester he contributed to the PUG project on genetically modified food, and ethnic marking of assisted reproduction across different European countries. His film, *The Way of the Road*, records Himalayan villagers' anticipation of a road to connect across the Tibetan border.

Cathrine Degnen is Lecturer in Social Anthropology at Newcastle University, UK. One of her primary research interests is how people grapple with social change in their everyday lives. This has led her to work on topics such as ageing and the self, memory and place, and food and genetic technologies. She is currently writing a monograph, *Years in* *the Making: Ageing Selves and Everyday Life in the North of England* for Manchester University Press, and has also published in journals such as *The Sociological Review* and *the Journal of the Royal Anthropological Institute*.

Penelope Dransart is Reader in Anthropology and Archaeology at the University of Wales, Lampeter. Her fieldwork in Isluga, northern Chile, since 1986 has examined how herders and weavers have shaped and sustained a pastoral way of life. She is author of *Earth, water, fleece and fabric: an ethnography and archaeology of Andean camelid herding* (2002). Currently she is writing a book based on an ethnography of colours, which explores people's use of colours in relation to the cultural and environmental contexts in which in textiles are produced.

Cristina Grasseni lectures in Cultural Anthropology at the Graduate School in Anthropology and Epistemology of the University of Bergamo, Italy. Her research interests include alpine heritage and local development, responsible innovation and the anthropology of food production. She has published in English, German and Italian, including four monographs and seven edited volumes, amongst which *Developing skill, Developing Vision: Practices of Locality in an Alpine Community* (Berghahn, 2009) and *Skilled Visions: Between Apprenticeship and standards* (ed., Berghahn 2007).

Sarah Green is Professor of Social Anthropology at the University of Manchester. Her research looks at how people locate themselves in the world and in relation to others, and particularly focuses on epistemology and the making and remaking of boundaries. She has conducted research among radical feminists in London, among information and communications technology (ICT) specialists in Manchester, and her current work examines the shifting borders of Europe around the Aegean. Her publications include *Notes from the Balkans: Locating marginality and ambiguity on the Greek-Albanian border.* (Princeton University Press, 2005) and *Urban Amazons: Lesbian Feminism and Beyond in the Gender, Sexuality and Identity Battles of London.* (Macmillan, 1997).

Richie Nimmo is Lecturer in Sociology at the University of Manchester. His research explores the ambiguous status of nonhumans in modern knowledge-practices, and the constitution of 'the social' across materially heterogeneous relations, systems and flows. He is the author of *Milk,*

Modernity and the Making of the Human: Purifying the Social, published by Routledge in February 2010.

Marcia Stephenson is Associate Professor of Spanish at Purdue University. Her book *Gender and Modernity in Andean Bolivia* (University of Texas Press, 1999) received the A.B. Thomas Award for Excellence. Other publications have appeared in journals such as the *Hispanic American Historical Review*, the *Latin American Research Review*, *Escarmenar: Revista Boliviana de Estudios Culturales*, *Chasqui*, *MLN, Revista de Estudios Hispánicos, Debate Feminista, Mallki,* and *Dispositio/n*. She is currently finishing a book manuscript with the title "Natural and Unnatural Histories of Andean Camelids in the Transatlantic World, 1568-1960."

Andrew Whitehouse is Teaching Fellow in Anthropology at the University of Aberdeen. His doctoral research was into relations between conservation organisations and farmers on the Scottish island of Islay. More recently he has conducted research into the ways in which people listen to birds and respond to their sounds.

INDEX